A Simple Story
of a
"Not-So-Simple" Universe

A Simple Story
of a
"Not-So-Simple" Universe

It's not the only story.
It's just my story!

Jerry Miller

Library of Congress Control Number:		2008911161
ISBN:	Hardcover	978-1-4363-8979-2
	Softcover	978-1-4363-8978-5

To order additional copies of this book, contact:
Xlibris Corporation
1-888-795-4274
www.Xlibris.com
Orders@Xlibris.com
54789

Contents

Preface...7

Introduction..9

Acknowledgments...11

Chapter 1:   The Basic Atom 13

Chapter 2:   Subatomic Particles..38

Chapter 3:   Important Fundamentals in Physics................................48
    1.   The Four Forces in Nature
    2.   Einstein's Mass/Energy Equation: (E = mc^2)
    3.   Quantum Mechanics
    4.   Relativity

Chapter 4:   The Big Bang 65

Chapter 5:   Earth, Life, and the End..79
    Birth of Earth
    The Evolution of Life
    The End of the World as We Know It

Chapter 6:   Strings—and Things...103

Chapter 7:   Game Changers of Science..113

References...119

Notes...121

Index...127

Preface

THIS IS A story of the universe as told in a most elementary way. It is my intent to include not only the microworld of the atom but also the macroworld of the far reaches of the cosmos. Along the way, we will include a stop at our planet Earth to review the processes of its geological birth and to study the evolutionary processes resulting in simple life-forms. Human life emerged roughly 13.7 billion years after the big bang and 4.4 billion years after the birth of earth. You will learn about the advances in science beginning in Greece in BC time and ending with the most recent theories of the universe. You will be amazed at the scientific breakthroughs throughout history.

The ancient Grecian philosophers first asked the question, What is the smallest piece of matter in the universe? Twenty-four hundred years later the atom's secrets were discovered, consisting of electrons, protons, and neutrons. Later in the 1960s, the protons and neutrons were found to consist of even smaller particles called quarks. Today there are hundreds of subatomic particles. We will learn that at the beginning of the universe, all of these subatomic particles were part of a primordial soup at extremely high temperature. As the universe expanded, it cooled allowing for subatomic particles to form into atoms such as hydrogen and helium. Hydrogen supplied the energy that fuels the stars even today. With further cooling, larger atoms formed, which are now the foundation of Earth and all of the other planets in the universe.

There are theories on the fate of our solar system and us on earth and ultimately on the universe. The story is immensely interesting not only from a scientific viewpoint but from the viewpoint of the future of all of humankind.

Introduction

Holding to the scientists' experiences as all important, I define universe, including both the physical and metaphysical, as follows: The universe is the aggregate of all of humanity's consciously-apprehended and communicated experience with the nonsimultaneous, nonidentical, and only partially overlapping, always complementary, weighable and unweighable, ever omni-transforming, event sequences. (*Operating Manual for Spaceship Earth*, R. Buckminster Fuller, 1969)

WITH ALL DUE respect to Buckminster Fuller, who was nominated for the 1969 Nobel Peace Prize, I haven't a clue as to what the above statement means. Perhaps its ambiguous nature carries a touch of levity regarding the state of the universe. Yet the nature of the universe in 1969 was probably thought to be better understood than it is today. Today with the continued help of mathematics and with the mounds of new data being generated by high-speed particle accelerators, and the Hubble space telescope, the knowledge gained has been astounding. But this knowledge comes with many more questions and surprises, taking aim at some of the bedrock fundamentals of science. Recently I heard the Discovery Channel make mention of a white hole (not a white dwarf) as the opposite of a black hole, and I asked myself aloud, "Where did that come from?" The state of the universe today is exciting and scary at the same time, and I can hardly wait to see where this story is going!

It is my objective to write a story of the universe, a story that will be interesting to read and simple to comprehend. As such, it will be elementary. I do this not only to satisfy self-commitment but to share with others who find the subject of the universe interesting on this level. It will not be easy to simplify the universe, but this is my challenge.

Acknowledgments

S PECIAL THANKS TO my daughter Tracey for her persistence that I complete this story and to my wife Patti for her encouragement and patience. Without the two of them, it would have been easy for me to forgo what has been my own life-long commitment.

Chapter One

The Basic Atom

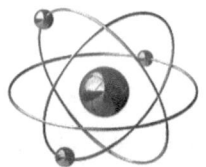

Early Greece (ca. 500 BC)

OF THE MANY Grecian philosophers in the time 500 BC, Leucippus (490 BC-?) is the first person known by name to consider the problem of splitting matter into fragments so small that they could no longer be split. He came to the conclusion that the splitting process could *not* go on forever and that a fragment of matter would remain. As simple as this may sound, it is quite a profound statement. This opinion differed from many other philosophers at the time who thought that the splitting process could go on forever. A student of Leucippus, Democritus (460-379 BC), accepted the notion of a remaining fragment and named it the "atomos" (meaning incapable of being cut). The concept of an atom was at this time purely philosophical for there was not a clue as to what the remaining fragment looked like. This was to wait for another 2,300 years.

By way of interest, two of the more notable detractors of atomism at the time were Aristotle (384-322 BC) and Plato (427-347 BC).

Fast Forward 2,000 Years to about 1500

By the early 1500s, an incredible 2,000 plus years beyond Democritus, no headway had been made between the believers and the nonbelievers of atomism. While atomism had been gaining in popularity, science had yet to produce any hard evidence for or against it. Nobody knew the inner workings of the atom. Consider that this is only 441 years from the advent of the first atomic bomb, yet no breakthroughs for the existence of the atom had been established.

The First Evidence of Atomism (1662)

So, atomism continued to gain in popularity through the 1600s. The first person to establish evidence for atomism was British scientist Robert Boyle (1627-1691). In 1662, he poured mercury into a transparent J-shaped tube with only the short end closed. He noticed a continual contraction in the short tube as mercury was poured more and more into the long tube. In fact, when the mercury poured into the long end was doubled, the contraction of air in the short tube decreased by half. This was interpreted as atoms in the air pocket being forced together as pressure increased. Since nobody could explain Boyle's results without the reasoning of atoms, this proved a turning point for atomism. However, at this point, there was still no explanation for the structure of the atom, and it would be another roughly 238-280 years before the final proof would sway all scientists.

The Search for Elements (1600s to 1800s)

The 1700s and 1800s would be eras of continuous discovery of the elements, characterization of their properties, development of the periodic table of the elements, and advancement of qualitative and quantitative techniques in chemistry. Since many elements appear in nature as compounds, such as ores and salts, chemists had to learn to separate the reacted products in order to isolate the pure elements. The primary gases, oxygen, hydrogen, and nitrogen were readily defined as were commonly available heavier elements such as iron, copper, tin, lead, silver, gold, aluminum, mercury, and others. The age of chemistry had begun.

Periodic Table Established (Mid-1800s to Early 1900s)

By 1809, at least 47 elements had been discovered, and scientists began to see patterns in their characteristics. In 1863, English chemist John Newlands divided the then discovered 56 elements into 11 groups based on their characteristics. And by 1869, Russian chemist Dmitri Mendeleyev started the development of the periodic table of the elements. This table arranged the elements by atomic mass. He predicted the discovery of other elements and left gaps in the table for those elements to be found. It is amazing that Mendeleyev was able to put the table together without any knowledge about the inner structure of the atom. Later in 1913, Henry Moseley showed that x-ray emission measurements of the elements coincided with the order of the elements by atomic number rather than atomic weight. By arranging the elements in order of atomic number, many of the early problems of the periodic table were solved.

Atomic number is defined as the number of protons in the nucleus, and *atomic mass* is defined as the total number of protons and neutrons in the nucleus.

Table 1
Periodic Table of the Elements

Legend (Groups with similar properties, Left to Right):
- Alkali Metals
- Alkaline Earth Metals
- Transition Metals
- Metals
- Metalloids
- Nonmetals
- Halogens
- Noble Gasses

(Key: Atomic Number / Element Symbol / Atomic Mass (Approximate))

1	2	3	4	5	6	7	8	9	10	11	12	13	14	15	16	17	18
1 H 1																	2 He 4
3 Li 7	4 Be 9											5 B 11	6 C 12	7 N 14	8 O 16	9 F 19	10 Ne 20
11 Na 23	12 Mg 24											13 Al 27	14 Si 28	15 P 31	16 S 32	17 Cl 35	18 Ar 40
19 K 39	20 Ca 40	21 Sc 45	22 Ti 48	23 V 51	24 Cr 52	25 Mn 55	26 Fe 56	27 Co 59	28 Ni 59	29 Cu 64	30 Zn 65	31 Ga 70	32 Ge 73	33 As 75	34 Se 79	35 Br 80	36 Kr 84
37 Rb 85	38 Sr 88	39 Y 89	40 Zr 91	41 Nb 93	42 Mo 96	43 Tc 98	44 Ru 101	45 Rh 103	46 Pd 106	47 Ag 108	48 Cd 112	49 In 115	50 Sn 119	51 Sb 122	52 Te 128	53 I 127	54 Xe 131
55 Cs 133	56 Ba 137	57 La 139	72 Hf 178	73 Ta 181	74 W 184	75 Re 186	76 Os 190	77 Ir 192	78 Pt 195	79 Au 197	80 Hg 201	81 Tl 204	82 Pb 207	83 Bi 209	84 Po 209	85 At 210	86 Rn 222
87 Fr 223	88 Ra 226	89 Ac 227	104 Rf 261	105 Db 262	106 Sg 266	107 Bh 264	108 Hs 277	109 Mt 268	110 Ds 269	111 Rg 272	112 Uub 277	113	114 Uuq 289	115	116 Uuh 292	117	118

Lanthanides:

58	59	60	61	62	63	64	65	66	67	68	69	70	71
Ce	Pr	Nd	Pm	Sm	Eu	Gd	Tb	Dy	Ho	Er	Tm	Yb	Lu
140	141	144	145	150	152	157	159	163	165	167	169	173	175

Actinides:

90	91	92	93	94	95	96	97	98	99	100	101	102	103
Th	Pa	U	Np	Pu	Am	Cm	Bk	Cf	Es	Fm	Md	No	Lr
232	231	238	237	244	243	247	247	251	252	257	258	259	262

Table 1. Periodic Table of the Elements

There are many variations of periodic tables of the elements on the internet. One version is shown on table 1. It provides you with the symbols of each of the elements along with their atomic numbers and their approximate atomic masses. It also provides the various groups of elements that have outer electron shells that are very much alike and therefore have chemical properties that are very much alike (more later). Note that elements 114 and 116 have just recently been discovered, and the elements 113, 115, 117, and 118 are still waiting to be discovered or are in the process of discovery. So to date, 114 elements have been discovered. Interestingly, as recently as 1991, at the time of Isaac Asimov's publication of *Atom,* there were only 106 elements, eight less than today. Remember, each element is represented by its own atom. The elements are listed in this periodic table by symbols that are not always helpful. For example, gold is represented as Au, potassium as K, lead as Pb, etc. Others, like lithium as Li, carbon as C, and nickel as Ni, etc., are readily obvious. To help you out, I have attached a listing of all of the elements alphabetically by full name and included their symbols, atomic numbers, and their atomic names. Refer to table 2. We will come back to the periodic table after we study the structure of the atom.

Atomic Number	Element	Symbol	Atomic Mass	Atomic Number	Element	Symbol	Atomic Mass	Atomic Number	Element	Symbol	Atomic Mass
1	Hydrogen	H	1.0079	26	Iron	Fe	55.845	51	Antimony	Sb	121.76
2	Helium	He	4.0026	27	Cobalt	Co	58.933	52	Tellurium	Te	127.6
3	Lithium	Li	6.941	28	Nickel	Ni	58.693	53	Iodine	I	126.904
4	Beryllium	Be	9.0122	29	Copper	Cu	63.546	54	Xenon	Xe	131.293
5	Boron	B	10.811	30	Zinc	Zn	65.39	55	Cesium	Cs	132.906
6	Carbon	C	12.0107	31	Gallium	Ga	69.723	56	Barium	Ba	137.327
7	Nitrogen	N	14.0067	32	Germanium	Ge	72.64	57	Lanthanum	La	138.906
8	Oxygen	O	15.9994	33	Arsenic	As	74.922	58	Cerium	Ce	140.116
9	Fluorine	F	18.9984	34	Selenium	Se	78.96	59	Praseodymium	Pr	140.908
10	Neon	Ne	20.1797	35	Bromine	Br	79.904	60	Neodymium	Nd	144.24
11	Sodium	Na	22.9897	36	Krypton	Kr	83.8	61	Promethium	Pm	145
12	Magnesium	Mg	24.305	37	Rubidium	Rb	85.468	62	Samarium	Sm	150.36
13	Aluminum	Al	26.9815	38	Strontium	Sr	87.62	63	Europium	Eu	151.964
14	Silicon	Si	28.0855	39	Yttrium	Y	88.906	64	Gadolinium	Gd	157.25
15	Phosphorous	P	30.9738	40	Zirconium	Zr	91.224	65	Terbium	Tb	158.925
16	Sulfur	S	32.065	41	Niobium	Nb	92.906	66	Dysprosium	Dy	162.5
17	Chlorine	Cl	35.453	42	Molybdenum	Mo	95.94	67	Holmium	Ho	164.93
18	Argon	Ar	39.948	43	Technetium	Tc	98	68	Erbium	Er	167.259
19	Potassium	K	39.098	44	Ruthenium	Ru	101.07	69	Thulium	Tm	168.934
20	Calcium	Ca	40.078	45	Rhodium	Rh	102.906	70	Ytterbium	Yb	173.04
21	Scandium	Sc	44.956	46	Palladium	Pd	106.42	71	Lutetium	Lu	174.967
22	Titanium	Ti	47.867	47	Silver	Ag	107.868	72	Hafnium	Hf	178.49
23	Vanadium	V	50.941	48	Cadmium	Cd	112.411	73	Tantalum	Ta	180.948
24	Chromium	Cr	51.996	49	Indium	In	114.82	74	Tungsten	W	183.84
25	Manganese	Mn	54.938	50	Tin	Sn	118.71	75	Rhenium	Re	186.207

Atomic Number	Element	Symbol	Atomic Mass	Atomic Number	Element	Symbol	Atomic Mass
76	Osmium	Os	190.23	101	Mendelevium	Md	258
77	Iridium	Ir	192.217	102	Nobelium	No	259
78	Platinum	Pt	195.078	103	Lawrencium	Lr	262
79	Gold	Au	196.966	104	Rutherfordium	Rf	261
80	Mercury	Hg	200.59	105	Dubnium	Db	262
81	Thallium	Tl	204.383	106	Seaborgium	Sg	266
82	Lead	Pb	207.2	107	Bohrium	Bh	264
83	Bismuth	Bi	208.98	108	Hassium	Hs	277
84	Polonium	Po	209	109	Meitnerium	Mt	268
85	Astatine	At	210	110	Darmstadtium	Ds	
86	Radon	Rn	222	111	Roentgenium	Rg	272
87	Francium	Fr	223	112	Uub		277
88	Radium	Ra	226	113			
89	Actinium	Ac	227	114	Uuq		289
90	Thorium	Th	232.038	115			
91	Protactinium	Pa	231.036	116	Uuh		292
92	Uranium	U	238.029	117			
93	Neptunium	Np	237	118			
94	Plutonium	Pu	244				
95	Americium	Am	243				
96	Curium	Cm	247				
97	Berkelium	Bk	247				
98	Californium	Cf	251				
99	Einsteinium	Es	252				
100	Fermium	Fm	257				

Table 2. The Elements in Order of Atomic Number

Finally, the Atom is Unveiled (1897 to 1932)

Now it is time to sidestep a bit for the purpose of reviewing the atom itself. For those of you who have not studied elementary chemistry, the last discussion of the periodic table probably did not help a lot. That's OK—consider it as history for now. We will return to it shortly.

All right! It is very late 1890s, and we still do not know what the atom is all about. We do know that each element has a different atom. It is now about 2,300 years from when Grecian Leucippus first conceptualized the atom as a "fragment."

Here goes! The following discoveries finally exposed the mysteries:

In 1898, English physicist J. J. Thomson first discovered electrons in the atom as small negatively charged particles. John Townsend and Robert Millikan determined their exact charge and mass.

In 1911, Ernest Rutherford and German physicist Hans Geiger discovered that electrons orbit the nucleus of an atom.

In 1914, Ernest Rutherford first identified protons in the atomic nucleus.

In 1931, James Chadwick first discovered neutrons in the atomic nucleus—this is only ten years from my birth date and fourteen years from the first atomic bomb.

Basics of the Atom

So in thirty-five years, from 1897 to 1932, the basics of the atom had been discovered and its little secrets exposed. The electrons are small, negatively charged, and orbit the nucleus of the atom. "Small" according to one reference claims that "electrons are point particles—they don't have any size at all." (*http://www.physlink.com/education/AskExperts/ae570.cfm*). Well, they are fundamental, but they do have mass and therefore size. The nucleus consists of protons and neutrons, both of which are much more massive than electrons. Protons carry a positive charge while neutrons have no electrostatic charge. Since the number of protons equal the number of electrons in each atom, atoms in there natural state are electrically neutral. We humans are, for the most part, electrically neutral (probably with the exception of the electrical synapses in the brain). Earth, the universe, and all celestial bodies are basically neutral.

Neutrons serve the purpose of stabilizing the repulsive nature of the protons in the nucleus basically to prevent the nucleus from exploding. The number of protons in the nucleus is called the atomic number, and the sum of protons and neutrons in the nucleus is called atomic mass. Since there are 114 different elements, there are 114 different atoms. Hydrogen is an exception with no neutrons in its nucleus. It simply has one proton. Helium exists naturally with two protons and two neutrons. Lithium, however, with an atomic number of three (three protons), requires four neutrons for stabilization. The greater the atomic number, the greater the number of neutrons required for stabilization. So that uranium with 92 protons requires 146 neutrons, and even then, the nucleus is unstable and is known as radioactive.

Relative Masses

Electron = 9.11 x 10^{-31} kg
Proton = 1.673 x 10^{-27} kg
Neutron = 1.675 x 10^{-27} kg

Scientists have actually measured these masses out to eight decimal points. Note that the neutron is slightly heavier than the proton. The electron is only 0.054% the mass of the neutron. The mass ratio of proton to electron is 1,836. One real life mass comparison has it that a 150-pound person consists of 149 lbs and 15 oz of protons and neutrons and only 1 oz of electrons.

How Small is Small?

It is very difficult to visualize or conceptualize how amazingly small the atom is. Even the numbers provided in the literature for size and mass although accurate are meaningless without some kind of perspective. Here are some attempts:

- In 1913, Jean Baptiste Perrin showed by his calculations that 100 million atoms side by side would be required to stretch out to 1 cm (less than ½ in).
- It would take over 100 million atoms to make a line as long as this—.
- It is known that the diameter of a nucleus is only 1/100,000th that of the atom—meaning that 99.999% of an atom is space!
- If gold were to be rolled into a sheet so thin that it would be transparent, it would still be 20,000 atoms thick.
- If the nucleus was the size of a grape, the electrons would be about 1 mi away on average, or if the nucleus was the size of a basketball, the electrons would be about 7 mi away. Everything else within the 7-mile radius would be *space*—nothing!
- Imagine that everything that seems to be opaque to the human eye is actually 99.999% space—the desk that I'm sitting at, your car, the

books on the shelf, a chunk of lead or a solid bar of gold. It's just that our eyes did not evolve to see that level of microscopic intrusion—to see that the atom is mostly space.

- How about a particle so small and so energetic that it can pass through trillions of miles of pure lead without hitting a single atom. It's called a neutrino. Crazy, isn't it?

Avogadro's Number

On the subject of size, this is a real-life example. With amazing insight, Amedeo Avogadro postulated (1766-1856) that the gram molecular weight of any element or molecule contains 6.022×10^{23} atoms. Now stay awake—this is chemistry talk, but it will be short. The element or molecule can be anything—water, sugar, gold, oil, protein, etc. Second, you need to know the molecular weight of the substance. If it is elemental just go to the periodic table of the elements and get the atomic mass. If it is a molecule, you need to know its molecular weight. So let's keep it simple and choose helium. The gram molecular weight simply means the atomic weight of helium in grams. From the periodic table, the atomic weight of helium is 4.0026. That number in grams contains 6.022×10^{23} atoms. Table salt is one atom of sodium to one atom of chlorine, expressed as the compound, sodium chloride. From the periodic table, one molecule contains 11 g of sodium and 17 g of chlorine (chloride). So by Avogadro's rule, 28 g of table salt contains 6.02×10^{23} atoms. Twenty-eight grams equals less than one-tenth of 1 lb of salt. Likewise, 55.8 g of iron, 32.1 g of sulfur, 29 g of copper, 92 g of uranium, etc., all contain 6.022×10^{23} atoms.

Mass Perspective

Let us pretend that we can isolate a small handful of neutrons and could pour them into a $1 \times 1 \times 1$ in^3 box until full. This is not altogether that weird since there are actually neutron stars out there. All we know right now is that protons and neutrons are much more massive than electrons, but what does that mean? Would the neutron-filled cube be as heavy as a bar of gold? If you are thinking much, much, much heavier, you are right. I will spare you some of the calculations, but I will give you enough information to do your own calculations if you are so inclined. We will assume that a neutron is a sphere (which isn't strictly so).

This is the equation:

Total mass of the neutrons inside of the box) = [(mass of one neutron) (volume of the empty box in cm³)] ÷ (volume of one neutron in cm³)

The mass of a neutron is 1.675×10^{-27} kg and the volume of the box is 16.4 cm³. For the volume of a neutron use the equation for a sphere which is $4/3\pi r^3$, and use the radius as 5.0×10^{-14} cm (from the Standard Model of Fundamental Particles and Interactions—assuming that the indicated value of 10^{-15} m is the diameter and not the radius). The result is roughly 4 thousand billion lbs in one in³. Do you find this incredible?

Can this be substantiated? One source from Website http://www.seasky. org/cosmic/sky7a08.html claims that a piece of neutron star the size of a sugar cube would weigh 100 million tons. By converting tons to pounds, we get 200 billion lbs. I would say that both examples qualify as mind-boggling. In case you are wondering, both masses would be much more than the crust of earth could hold and would crash right through to earth's center or continue on to the nearest black hole.

We know that electrons are not very massive and are viewed by some as "point" particles without mass to speak of. Yet they do have mass— roughly 2,000 times less than the masses of neutrons and protons. For the fun of it let's apply the previous box calculation to a hypothetical electron-filled box (would never happen because of electrical repulsion). We will use 5×10^{-19} cm for the radius of the electron and of 9.1×10^{-31} kg for the mass. This calculation comes out to be roughly 4 thousand billion billion lbs/in³, certainly enough to play a role in the total mass count of the universe.

Electron Orbits

Electrons revolve around the nucleus in very discrete orbits or shells. The inner orbit is labeled s and succeeding orbits are p, d, f, g, h and j (sometimes labeled K, L, M, N, O, P, and Q). If we assign a number to the shells rather than letters, we can calculate the maximum number of electrons allowed in each shell. The calculation is $2n^2$, where n = shell number. So shell 1 = 2×1^2 or 2 electrons, shell 2 = 2×2^2 or 8 electrons, shell 3 = 2×3^2 or 18 and so on with 32, 50, 72, and 98 being maximum for shells 4, 5, 6, and 7, as shown on illustration 1. In reality, this drawing would be a three-dimensional sphere with their electron orbits portrayed as discrete, clouded shells.

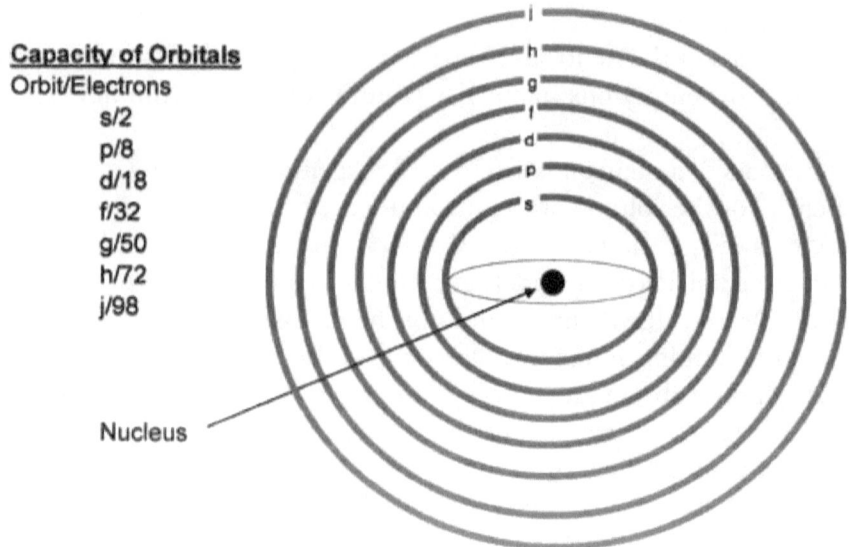

Capacity of Orbitals
Orbit/Electrons
s/2
p/8
d/18
f/32
g/50
h/72
j/98

Nucleus

Illustration 1. Electron Shells

Table 3 is the same periodic table as table 1, with the exception that within each element box is the actual electron code for each of the elements. You can see that elements 1 through 10 have 1 through 10 protons and 1 through 10 electrons, respectively. Hydrogen is 1, and neon is 10. Any element number that you select from the table will have that many protons and that many electrons. Uranium is element 92, so it has 92 protons and 92 electrons.

So again, elements 1 through 10 gradually fill the first two orbits—2 in the s and 8 in the p. Sodium (element 11) begins the process of filling shell d with one electron. By the time we get to argon (element 18), there are 8 electrons in the d shell, with none in the f shell. Now the coming sequence will place 2 electrons in the f shell, before coming back to fill the d shell to its maximum 18. So elements 19 and 20 (potassium and calcium) have 1 and 2 electrons in the f shell. The next ten elements (scandium through zinc) continue to fill the d shell to capacity of 18. So zinc finishes with a 2s, 8p, 18d, and 2f configuration—and so on through the 112 elements shown on table 3.

Table 3
Electron Configurations for all of the Elements

Element	Z	Electron Configuration
H	1	1
He	2	2
Li	3	2,1
Be	4	2,2
B	5	2,3
C	6	2,4
N	7	2,5
O	8	2,6
F	9	2,7
Ne	10	2,8
Na	11	2,8,1
Mg	12	2,8,2
Al	13	2,8,3
Si	14	2,8,4
P	15	2,8,5
S	16	2,8,6
Cl	17	2,8,7
Ar	18	2,8,8
K	19	2,8,8,1
Ca	20	2,8,8,2
Sc	21	2,8,9,2
Ti	22	2,8,10,2
V	23	2,8,11,2
Cr	24	2,8,13,1
Mn	25	2,8,13,2
Fe	26	2,8,14,2
Co	27	2,8,15,2
Ni	28	2,8,16,2
Cu	29	2,8,18,1
Zn	30	2,8,18,2
Ga	31	2,8,18,3
Ge	32	2,8,18,4
As	33	2,8,18,5
Se	34	2,8,18,6
Br	35	2,8,18,7
Kr	36	2,8,18,8
Rb	37	2,8,18,8,1
Sr	38	2,8,18,8,2
Y	39	2,8,18,9,2
Zr	40	2,8,18,10,2
Nb	41	2,8,18,12,1
Mo	42	2,8,18,13,1
Tc	43	2,8,18,14,1
Ru	44	2,8,18,15,1
Rh	45	2,8,18,16,1
Pd	46	2,8,18,17,1
Ag	47	2,8,18,18,1
Cd	48	2,8,18,18,2
In	49	2,8,18,18,3
Sn	50	2,8,18,18,4
Sb	51	2,8,18,18,5
Te	52	2,8,18,18,6
I	53	
Xe	54	2,8,18,18,8
Cs	55	2,8,18,18,8,1
Ba	56	2,8,18,18,8,2
La	57	2,8,18,18,9,2
Hf	72	2,8,18,32,10,2
Ta	73	2,8,18,32,11,2
W	74	2,8,18,32,12,2
Re	75	2,8,18,32,13,2
Os	76	2,8,18,32,14,2
Ir	77	2,8,18,32,15,2
Pt	78	2,8,18,32,17,1
Au	79	2,8,18,32,18,1
Hg	80	2,8,18,32,18,2
Tl	81	2,8,18,32,18,3
Pb	82	2,8,18,32,18,4
Bi	83	2,8,18,32,18,5
Po	84	2,8,18,32,18,6
At	85	2,8,18,32,18,7
Rn	86	2,8,18,32,18,8
Fr	87	2,8,18,32,18,8,1
Ra	88	2,8,18,32,18,8,2
Ac	89	2,8,18,32,18,9,2
Rf	104	2,8,18,32,32,10,2
Db	105	2,8,18,32,32,11,2
Sg	106	2,8,18,32,32,12,2
Bh	107	2,8,18,32,32,13,2
Hs	108	2,8,18,32,32,14,2
Mt	109	2,8,18,32,32,15,2
Ds	110	2,8,18,32,32,17,1
Rg	111	2,8,18,32,32,18,1
Uub	112	2,8,18,32,32,18,2
	113	
Uuq	114	
	115	
Uuh	116	
	117	
	118	

Lanthanides

Element	Z	Electron Configuration
Ce	58	2,8,18,20,8,2
Pr	59	2,8,18,21,8,2
Nd	60	2,8,18,22,8,2
Pm	61	2,8,18,23,8,2
Sm	62	2,8,18,24,8,2
Eu	63	2,8,18,25,8,2
Gd	64	2,8,18,25,9,2
Tb	65	2,8,18,27,8,2
Dy	66	2,8,18,28,8,2
Ho	67	2,8,18,29,8,2
Er	68	2,8,18,30,8,2
Tm	69	2,8,18,31,8,2
Yb	70	2,8,18,32,8,2
Lu	71	2,8,18,32,9,2

Actinides

Element	Z	Electron Configuration
Th	90	2,8,18,32,18,10,2
Pa	91	2,8,18,32,20,9,2
U	92	2,8,18,32,21,9,2
Np	93	2,8,18,32,22,9,2
Pu	94	2,8,18,32,24,8,2
Am	95	2,8,18,32,25,8,2
Cm	96	2,8,18,32,25,9,2
Bk	97	2,8,18,32,26,9,2
Cf	98	2,8,18,32,28,8,2
Es	99	2,8,18,32,29,8,2
Fm	100	2,8,18,32,30,8,2
Md	101	2,8,18,32,31,8,2
No	102	2,8,18,32,31,8,2
Lr	103	2,8,18,32,32,9,2

Table 3. Periodic Table with Electron Configurations for all of the Elements

Table 4 shows that only the first four orbits get filled to capacity with a configuration of 2s, p8, d18, and f32. This begins with ytterbium, element 70. All of the other higher elements, 71 to 118, also have these same four orbits filled to capacity. Orbit g has a capacity of 50 electrons but tops off with nobelium at 32. Orbit h has a capacity of 72 but tops off at 18 with roentgenium, atomic number 111. And orbit j has a capacity of 98 but tops off at 2 with radium, atomic number 88.

Table 4
Electron Shell Capacity

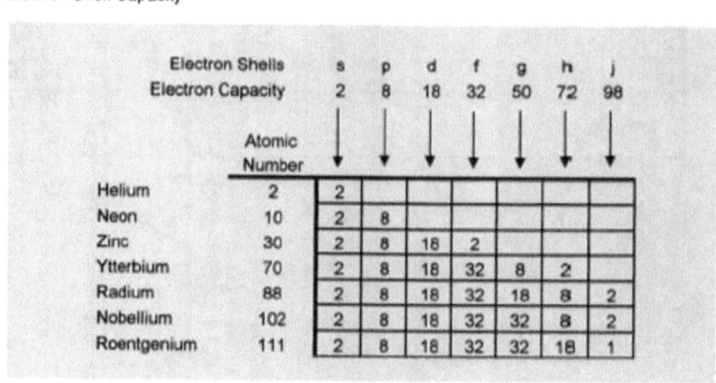

	Electron Shells	s	p	d	f	g	h	j
	Electron Capacity	2	8	18	32	50	72	98
	Atomic Number							
Helium	2	2						
Neon	10	2	8					
Zinc	30	2	8	18	2			
Ytterbium	70	2	8	18	32	8	2	
Radium	88	2	8	18	32	18	8	2
Nobellium	102	2	8	18	32	32	8	2
Roentgenium	111	2	8	18	32	32	18	1

Table 4. Electron Shell Capacity

Isotopes and the Atomic Nucleus

Within the nucleus, the protons are always constant for any given element—hydrogen is 1, helium is 2, etc. However, the number of neutrons can vary considerably and are always equal to or more than the quantity of protons. The only exception is the standard element, hydrogen, which has one proton and no neutron. However, most elements have one or more varieties differing in the total count of neutrons. These are called isotopes. Let's look at some real examples. Hydrogen has an atomic number of 1 (1 proton) and an atomic mass of 1 (0 neutrons). However, it has two isotopes, one with 1 neutron and another with 2 neutrons called deuterium and tritium, respectively. These isotopes are rare. Deuterium occurs in nature at less than 1 for every 6,000 hydrogen atoms, and tritium occurs at much less than that. Most elements have more than 1 isotope. Tin has the record with 10 stable isotopes. Approximately 270 stable isotopes and more than

2,000 unstable isotopes exist. For the most part, the isotopes simply exist as part of the parent element. Some like deuterium can be isolated and used in nuclear fuel applications.

The impact that isotopes have on the periodic table should be clarified. Note from the periodic table 1 that the atomic numbers of the elements are whole numbers, indicating the whole number of protons in the nucleus. One could anticipate that the atomic masses would equally add up to whole numbers if measured in atomic mass units (amu), yet you can see from table 2 that the atomic masses are shown as a decimal number (the decimals were rounded off on the periodic table, table 1). This is because the mass number reported is an average weight of a population of atoms that include all of their isotopes, rather than an individual atom. The weight then of a population of atoms that would include all of its isotopes is somewhat smaller or slightly bigger than the whole.

CAUTION

RADIATION AREA

Radioactivity

Let us change gears and go back to the nucleus. First, let's review the nucleus. Recall that the nucleus consists of protons and neutrons only. The number of protons always equals the number of electrons. Protons are actually composed of two quarks (to be studied later), which are considered elemental, and neutrons, which are composed of a proton and an electron. I hope that you are thinking, how did an electron get into a nucleus? We will get into that shortly. The electrons are considered elemental. The number of neutrons is equal to or greater than the number of protons. The larger the number of protons, the greater the number of neutrons required to stabilize the repulsion forces of the protons.

Alpha Decay

So with that, let's talk about radiation. This is important for a couple of reasons and ties into the sneaky electron getting into the nucleus. By definition, *radiation* is the spontaneous disintegration of a large unstable nucleus like radium or uranium to a more stable configuration by transforming into another element. Let us consider the nucleus of uranium. It has 92 protons and 146 neutrons in its nucleus. Since it is radioactive (unstable), an atom within uranium spontaneously emits an alpha particle, which is the same as a helium nucleus, 2 protons and 2 neutrons. This rather massive particle can be stopped by a sheet of paper and will not penetrate skin. The net result is that the atomic number of that atom decreases from 92 to 90 (two protons ejected), and the atomic mass decreases from 238 to 234 (2 protons and 2 neutrons)—the new atom is no longer uranium! A quick check of the periodic table for atomic number 90 indicates that the new atom is thorium, with an atomic mass of 234. Checking table 1 (which shows atomic mass numbers) shows that the standard thorium has an atomic mass of 232. So thorium 234 is an isotope of thorium 232 and is also radioactive. It is, however, one-step closer to stability.

Beta Decay

Thorium now goes through beta decay. Beta decay also occurs within an unstable nucleus. When it occurs, an electron is emitted. Recall this electron comes from a neutron in the nucleus, which consists of a proton and an electron. When this happens, the proton remains from what was a neutron, thus adding another proton to the nucleus and changing the atomic number to one higher. Actually for correctness, an antineutrino is also emitted. The beta particle can be stopped by 1-2 in of wood. The atomic mass remains the same because a neutron is changed into a proton.

So getting back to the above, where we have thorium 234/90 (atomic mass/atomic number) from the alpha decay of uranium 238/92, this thorium is also radioactive and goes through beta decay. So 1 proton is added to the nucleus, resulting in another element change to atomic number 91, protactinium 234/91. This decay goes on until the nucleus is stable at atomic number 82—lead, the highest stable element.

Gamma Decay

This is yet a third mechanism for radioactive decay. A gamma ray is a high-energy photon that originates from the nucleus as a result of a rearrangement of the electrical charge. This rearrangement occurs within heavy atoms that are radioactive, going through alpha and beta decay and changing the balance of protons and neutrons. Gamma rays are a form of electromagnetic radiation normally emitted from the electrons changing energy levels. Gamma rays neither change the atomic number nor the atomic mass of the nucleus when it is emitted. Gamma rays are the most penetrating form of radiation.

So radioactive decay results in transforming one element into another by alpha or beta decay. Gamma decay does not alter the originating atom. The rate at which this happens is measured in half-lives of the radioactive element, which is the time for 50% of a radioactive element to change into a new element. If for example the half-life of an unstable element is 10,000 years for 50% loss, then in another 10,000 years, the original element has been reduced to 25%. Unstable isotopes begin at atomic number of 83 (bismuth) and goes through 118. That range represents an extremely broad spectrum of half-lives. In fact, half-lives may vary from fractions of a second to billions of years. If you find this of interest, a listing of the half-lives of about 250 of the most common radioactive isotopes can be found at http://www.iem-inc.com/toolhafr.html.

Back to the Periodic Table

Note from table 3 that the periodic table has eighteen columns. First, let us look at column 18 (left to right) because the elements in that column are called the noble gases. Why noble? Well, notice that the outer orbit of each element is filled with a stable number of electrons. In the case of helium, the stable capacity is two. In the case of the higher elements in the same column, 8 electrons constitute a very stable state. Stable here means that these atoms do not require the assistance of other atoms to achieve stability. As such, they do not react well with other elements. Therefore, only a handful of compounds exist with any of the noble gases. Being "offish," they earn the name "noble."

Now let's jump back to column 1, where all of the elements are characterized by having 1 electron in the outer orbit. Because all of the elements "search" for stability, the elements in column 1 are attracted to the elements in column 17. If you're already looking at column 17, you are seeing that their outer orbit consists of 7 electrons, 1 short of stability. By giving up its outer electron, the elements in column 1 revert to a stable 8 electrons in the new outer orbit (or 2 in the case of lithium), and the atom is left with a net +1 charge. The atoms in column 17 gain the electron, thus filling the outer orbit with 8, and the atom is left with a net—1 charge. Thus, we have a win/win situation. Stable atoms of two different elements are bonded together by electrical attraction. And I'm sure you will recognize HCl (hydrochloric acid), NaCl and KCl (table salt)—see illustration 2. Similarly, elements from column 2 have 2 electrons in the outer orbit and have a natural affinity for elements in column 16, which have 6 electrons in the outer orbit. By giving up 2 electrons, atoms from column 1 have a full outer orbit, and atoms from column 16 in gaining 2 electrons are satisfied with a full outer orbit. So, atoms with 1, 2, or 3 electrons in their outer orbit tend to lose them in interactions with atoms that have 5, 6, or 7 electrons in their outer orbit and vice-versa. Atoms that have 4 electrons in the outer orbit will tend to share. Thus, we see the basis for chemistry and the value of the periodic table.

Hydrochloric Acid (HCl)

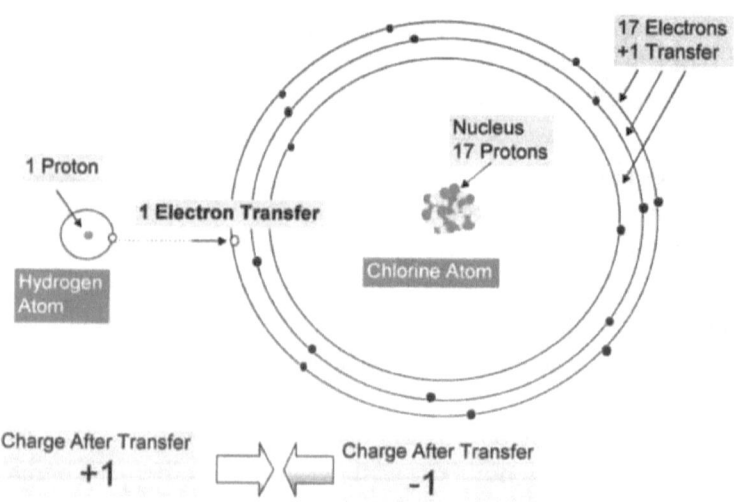

Illustration 2. Atoms Giving and Sharing Electrons

With an understanding now of the atomic nucleus and electrons, it will be easier to make some sense out of the periodic table on table 1. First, allow me to provide some comments. The elements are arranged in increasing order of atomic number in rows from left to right. Recall atomic number equals the number of protons in the nucleus and go up to 118, four of which have yet to be discovered. We discussed columns 1, 2, 17, and 18 with elements in each column having similar physical and chemical properties. In addition, there are nonmetals, consisting of carbon, nitrogen, oxygen, fluorine, phosphorous, sulfur, and selenium in columns 14, 15, and 16, and seven *metalloids*, which include boron and silicon. There are seven metals, which include aluminum, tin, and lead. And there are thirty-six transition metals from columns 3-12, which include chromium, iron, nickel, zinc, silver, platinum, gold, mercury as the more common ones. Elements 57-71 are usually called rare earth elements since they are "rarely" found on earth and exist only in very small quantities. Elements 89-103 include most of the elements that are found in nuclear reactions and include uranium and plutonium. Noteworthy elements larger than atomic number 92 (uranium) do not occur naturally. They have all been produced artificially through high-speed particle bombardment.

The beauty of the periodic table is that it is the bedrock of chemistry, not only for earth but for the chemistry of the entire universe. Shortly after the big bang, matter existed as a super-hot cauldron of subatomic particles. Atoms did not exist yet. Once the atoms formed, they became the building blocks for all of the matter in the universe. How are we sure? Well, by the periodic table, the nucleus of the atom contains protons and neutrons. The number of protons in the nucleus, as we know, is the atomic number and is a whole number. There is no such thing as a half-proton or a quarter-proton. Therefore, we have hydrogen at 1, helium at 2 and up to uranium at 92 elements occurring naturally on earth.

Functions of Electrons
- The electron cloud repulsion around atoms keeps them from crashing into each other. This ensures that atoms always maintain their vast amount of space. This is universally true except for neutron stars, white dwarfs, and black holes, which are immensely dense because the electrons have been pushed into the nucleus where they combine with the protons to become neutrons.
- Electrons enter into chemical reactions with other elements—the nucleus does not. Some elements can lose an electron leaving the atom with a net positive charge; some can gain an electron leaving the atom with a net negative charge, and others can share electrons and maintain neutrality. The drive in favor of reactions involves stabilizing the outer orbit with, for example, 2 or 8 electrons.
- Electrons can absorb and emit electromagnetic radiation in the form of energy depending upon outside influences. This is the basis of visible light, ultraviolet radiation, infrared radiation, radio waves, and others. Each electron orbit represents an energy level: the closer to the nucleus, the lower the energy level.
- Electrons are responsible for electricity, AC and DC. We all remember that famous experiment in school where static electricity was created by rubbing a glass rod with a piece of amber.

- Electrons are thought to be "elemental" unlike neutrons and protons and cannot be further divided. Remember that a neutron is made up of a proton and electron, and a proton is made up of quarks, which are considered elemental (later subject).

Electron and Atomic Spacing

For this example, think of a carbon atom with a nucleus of 6 protons the size of a basketball. We know that the 6 electrons would be circling at a radius of about 7 mi and would be circling the nucleus as a spherical cloud. None of the space between the nucleus and the electrons (7 mi) can be violated by another atom because their electron clouds would repel each other. I do not know how far the electron clouds in the two atoms need to be for a stabilized condition, but it probably depends upon the type of element—number of electrons, etc. One might naturally ask the question, "can't air get between the nucleus and the electron cloud?" Well, in the case of 7 mi, that would seem plausible. However, this is just a hypothetical blown-up illustration. We need to scale-up the air atom to the same size as the hypothetical illustration, and when we do that and visualize the oxygen atom right next to the carbon atom, we see that a straight line from one nucleus to the other is now 14 mi, with absolutely nothing in between—space!

Electrons and Electromagnetic Radiation

Electromagnetic radiation is not as scary as it sounds. And since it is very important to us human beings in all aspects of everyday life, we need to spend some time here. Recall that electrons circle the nucleus of the atom in discreet shells. In 1913, the Danish physicist Niels Bohr proposed that electrons occupy specific shells with very specific energy levels ranging from low to high. The electrons nearest to the nucleus are the lowest in energy. Bohr proposed that electrons cannot move freely from one shell to the next but could move in precise steps when energy, such as heat, is applied. Electrons subjected to heat (infrared radiation) absorb the energy and jump to a higher shell—the next outer shell. When the electrons fall back to the lower level, they release a precise quanta of energy in the form of specific wavelengths of electromagnetic radiation. There are many scientific principles implicit in that statement.

Quanta are photons! Electromagnetic radiation is a stream of photons. Now the very word *photon* wants to make us think that it is a particle—like electron, proton, and neutron. They are not, even though some scientists refer to them as "massless particles." This may be confusing to all of us to use the word *particle* for something that is not a particle. Other terms for photons are "bundles of energy" or "quanta." Electromagnetic radiation is responsible for radio waves, microwaves, infrared rays, visible light, ultraviolet rays, x-rays, and gamma rays. Again this is the same gamma ray that can also originate from the nucleus as radioactive decay. Visible light, well, is what we see: the sun, incandescent light bulbs, fluorescent light bulbs, candles, television, lightning, etc.

Refer to illustration 3. Visible light is only a tiny portion of the entire electromagnetic radiation spectrum. Consider that we see only the wavelength range of 0.00007 cm (color red) to 0.00004 cm (color violet). This is the range of visibility; any wavelength out of this range is invisible to the eye. The invisible portions of the spectrum also consist of photons but differ in wavelength. The only difference in the photons that we see and that we don't see are the energy levels implicit in the photons.

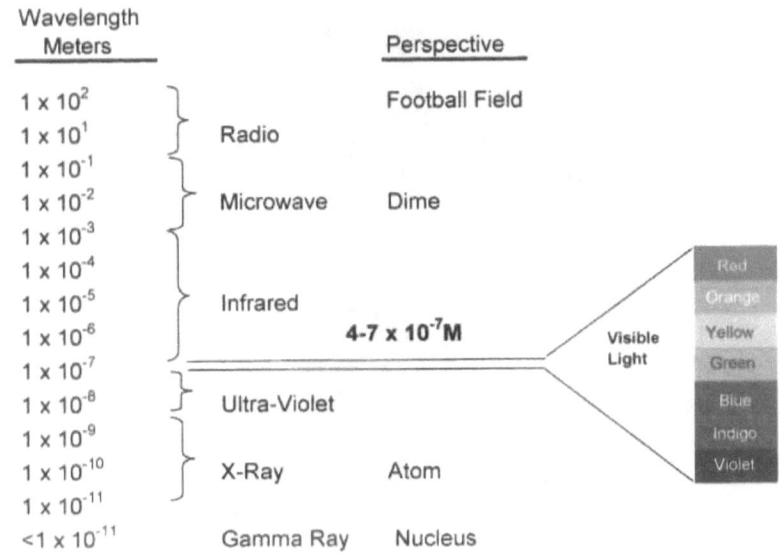

Illustration 3.

The Electromagnetic Spectrum

In 1665, Isaac Newton produced his first rainbow in a darkroom by shining a light source through a glass prism. He noticed the order of colors as red, orange, yellow, green, blue, indigo, and violet. Just as interesting is when he placed another prism in the path of the diffracted colors, they returned to white light. This created a problem in understanding that particles theories did not satisfy. At about the same time, Christiaan Huygens formulated a wave theory that seemed to better explain the prism rainbow. It wasn't until 1801 that English physicist Thomas Young performed a crucial experiment. He set up a light source behind a vertical surface with two vertical slits and behind that a vertical screen. Young reasoned that if light consisted of particles, "the overlapping region should receive particles from both slits and be brighter than the outlaying regions that received particles from only one slit or the other." But it was not so. Young continues with "The wedge of light from each slit fell on a screen and overlapped, resulting in a pattern of stripes—bright bands and dim bands alternating." Light indeed consisted of waves. (*The Atom, Isaac Asimov, p. 29*)

Electricity

We already know the foundations of electricity with the knowledge that opposite charges attract, and like charges repel. Electricity, however, is about moving electrons. The attractive electrical force between an electron and a proton is expressed by a mathematical formulation that is very similar to that of gravity (which is not an electrical attraction). Just by way of information, the charge on a single electron and a single proton—their charges are equal—is expressed as 1.6×10^{-19} coulombs. All ordinary objects are electrically neutral.

Static Electricity

Static electricity implies that some electrons are stored as potential energy. If the electrical charge builds up to an overload, it will eventually discharge in the form of a spark to return to neutrality. Electrons in many cases are easy to remove from an atom, particularly those in the outer shells. Rubbing a balloon on your hair will make your hair stand up. Why? Electrons rubbed on to your hair from the balloon provide a negative charge to your hair. Your individual hairs are repelling each other. Another example I so vividly remember is from my experiences in the northern states during wintertime. With little moisture in the air, static charge built up until it, well, discharged, resulting in an alarming spark and a small but effective, electrical jolt. This served as an unwanted wake-up call on many—a cold, blustery morning. Another example is lightning, which is simply nature's way of discharging electrons on a grand scale from cloud to cloud and from cloud to ground.

Electrical Current

By definition, to move electrons through a solid, it is required that the solid be a conductor. Insulators do not work. The conducting wire must be connected to a battery or generator to create an electrical field. Why do metals act as good electrical conductors? This is so fundamentally basic, and it makes such sense when you hear it—metals in general do not hold firmly on to their outermost electrons. When pushed, as in electrical field, copper electrons, for example, are free to flow atom to atom. Conversely, insulators hang on to their outermost electrons very tightly.

Magnetism

This comes from a website (refer to Notes), and the article is titled "Ridiculously Brief History of Electricity and Magnetism." "In 900 BC, Magnus, a Greek shepherd, walks across a field of black stones, which pulls the iron nails out of his sandals and the iron tip form his shepherd's staff. This region became known as Magnesia."

In the eleventh century, it was discovered that if a needle made out of magnetic ore or steel that had been magnetized by stroking with magnetic ore was allowed to turn freely, it would always align itself north and south, with the north end always turning to the north. In 1600, William Gilbert, physician to Queen Elizabeth discovered that earth is a giant magnet. In 1784, the French physicist Charles Augustin de Coulomb (recognize the last name?) measured the force (either attraction or repulsion) between a magnetic north pole and a magnetic south pole and came up with the inverse square law—force declined as the square of the distance. This is similar to what Newton developed as the force of gravitational attraction in 1687. However, the magnetic strength of two needles is trillions of trillions of trillions times as strong as the gravitational strength of the same two needles. (*The Atom,* Isaac Asimov, p. 37). Most of you have been involved at some point in your lives in a laboratory experiment where iron filings were scattered in a field around a magnet. Amazingly, the invisibility of magnetism is exposed. We can see the actual magnetic lines of force. It is expected that earth's electromagnetic lines of force extend deeply out into space. Every atom in the universe is a magnet with a negative and a positive end. But the magnets are randomly turned every way. There are as many repulsions as there are attractions. Overall, the universe is left with the magnets canceling each other out.

So what then is the value of magnetism? Well, on earth, the combination of magnetism and electricity has given mankind a way to generate electricity and to transport it to billions of people. I am not going to go through its many, many, many benefits when all we need to do is count the electrical outlets in your homes and garages. We are going a little beyond atoms here, but it is important to understand how all of this happened.

In 1819, a Danish physicist by the name of Hans Christian Oersted (1777-1851) was lecturing his class on electric current. He held a compass near a wire with electric current. To his amazement, the needle on the compass immediately turned 90° to the direction of the current. When Oersted reversed the flow of electricity, the compass reversed 90° the other direction. He is credited as the first to demonstrate a relationship between electricity and magnetism. In 1820, a French physicist, André Marie Ampère showed that two live wires running in the same direction attracted each other and in opposite directions repelled each other. Current in itself created magnetism! If a live wire is twisted into a spring configuration (to be called a solenoid), each curve attracts the other, setting up a magnetic field. Now we are getting real close. In 1831, Faraday turned mechanical energy into electricity. He set up a system whereby a copper disk turned between the poles of a magnet. As long as the disk was turned, an electric current was induced. That same year, an American physicist by the name of Joseph Henry invented the electric motor. And the rest, as they say, is history.

Fission

There are two more subjects before moving on to subatomic particles—atomic fission and atomic fusion. The two processes are very much different but are similar in that they release a huge amount of energy. In terms of fission, there is a destructive form called the atomic bomb and a constructive form called nuclear power.

So what is fission? *Fission* is the splitting of a nucleus of a large atom by adding a neutron to a nucleus that is already unstable (radioactive) due to its already high number of protons and neutrons in the nucleus. The added neutron splits the nucleus into two parts of lower atomic masses. In the process, a large amount of energy is released in the form of electromagnetic radiation (protect your eyes) and kinetic energy. In addition, two free neutrons are released. The free neutrons can then carry on the process to other atoms triggering four other neutrons into what is known as a chain reaction, splitting other nuclei at exponential speed. In terms of a bomb, it would be considered an out-of-control reaction (with a controlled outcome), and in terms of a nuclear reactor, it would be called a controlled reaction.

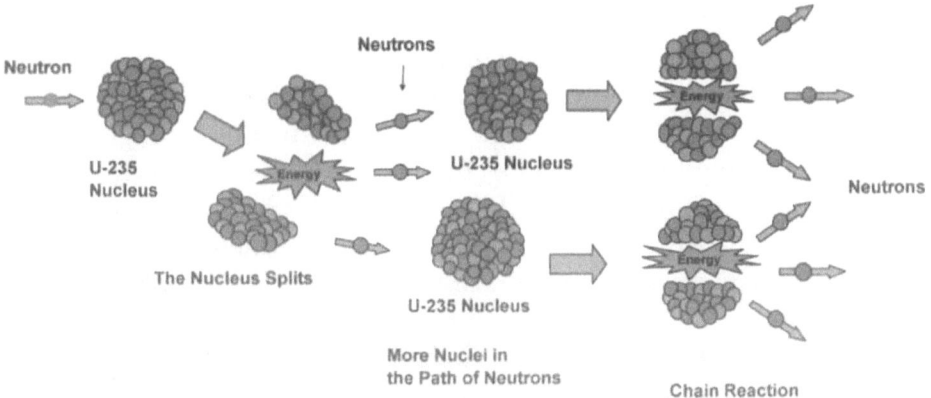

Illustration 4. Atomic Fission

Typical chemical isotopes that can sustain a chain reaction are uranium 235 (atomic mass) and plutonium 239 (atomic mass). They can be triggered by bombarding with neutrons (induced fission). The energy that is released in a fission reaction can be calculated by Einstein's famous equation $E=mc^2$ (energy = mass x velocity of light squared). Since the total rest mass of the fission products from a single reaction is less than the mass of the original fuel, nucleus mass was lost to energy. A slight loss in mass amounts to huge gains in energy. Typical fission events release about 200 million eVs (electron volts) of energy for each fission event. In contrast, most chemical reactions like TNT release at most a few eVs per event.

See illustration 4. Uranium 235 and plutonium 239 are the most common nuclear fuels. If we use uranium 235 as an example of fission, we know from the periodic table that uranium has an atomic number of 92, which means 92 protons. Since the balance is neutrons, there are 143 in the nucleus. It is an isotope of uranium 238, which has 146 neutrons and is the common form of uranium. So when fission of uranium 235 is induced by adding a neutron to the nucleus, it interferes with the strong nuclear force holding the nucleus together, and the nucleus splits into two smaller elements. There are a number of pathways that the splitting can take, but they must satisfy the rule that the sum of the protons in both elements equals 92, and that the total neutrons both in the nucleus and those liberated as free neutrons equals 236 (235 plus the inducing neutron). Two pathways for the proton rule are xenon 54 and strontium 38, kryptonite 36, and barium 56.

Fusion

Fusion is different than fission in two regards: (1) fusion starts with the least massive (and most abundant) element in the universe—hydrogen, rather than with very massive nuclei like uranium, and (2) fusion achieves its awesome power as evidenced by the hydrogen bomb by fusing two hydrogen atoms together, rather than by splitting massive nuclei. Sounds simple? Perhaps! The one principle however that remains common to both fission and fusion is that mass is turned into energy, and by Einstein's equation, it doesn't take a whole lot of lost mass to generate a massive amount of energy. If you are thinking about the atmosphere suddenly turning into hydrogen fireballs, you can rest your fears. It takes some very special conditions to fuse two hydrogen atoms. If you are thinking that it must be difficult to bring two atoms of any element together because of electron repulsion, you are right. And the trick to fusion is getting them together. And to do that means overcoming not only the repulsion of the electrons but of the positive charges of nuclei to nuclei for them to begin to feel the *strong nuclear force* (to be discussed in chapter 3). In fact, the two nuclei need to get close enough so that the strong nuclear force will pull them together into single nuclei. Refer to illustration 5. So, how do scientists get around these obstacles?—by attaching an atomic bomb!

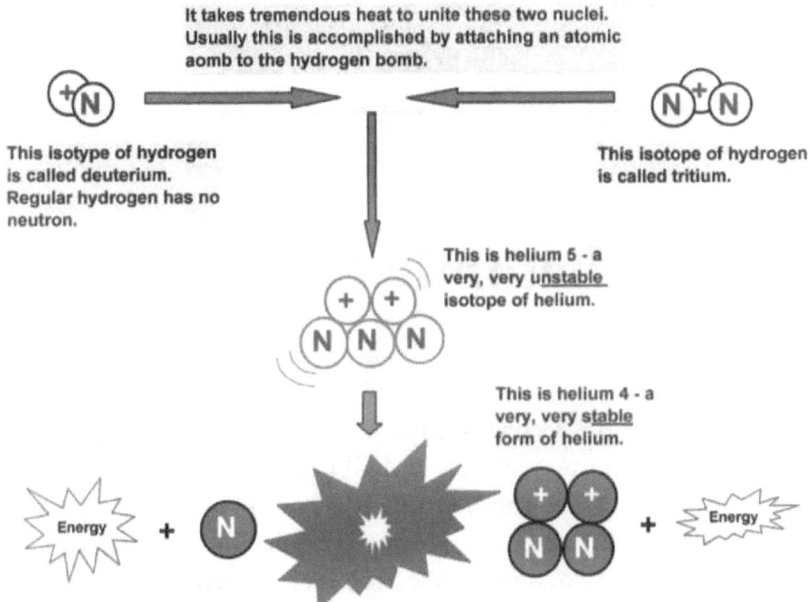

Illustration 5. Atomic Fusion

Interestingly, when the universe was cooling down, billions of years after the big bang and the elements were forming from hydrogen and helium up to much more massive elements, the fusion process was responsible for forming elements up to the element iron. Any elements higher in atomic weight than iron would have been formed by the fission process.

It has been mentioned earlier that fusion is the stuff of the stars, including our sun, and has been for billions and billions of years. Without the heat from our sun, life on earth would not have evolved. It would be a cold ice ball.

Chapter Two

Subatomic Particles

History

JUST FOR THE record, I choose to define electrons, protons, and neutrons as atomic particles and all others from within the atom as subatomic. Actually, since the neutron is a combination of a proton and an electron, we could reduce the atomic particles to just a proton and an electron. And of these two, only the electron is elemental. But we will see later that things are a little more complicated than that. It has been stated in the preface that there are now approximately 100 of these subatomic particles. It should be interesting trying to simplify this subject. Let's give it a go.

First, let's have a historical perspective. By 1931, the three primary particles had been discovered: the electron, the proton, and the neutron. But we have to back up one year for the first two subatomic particles, one of which is the photon and the other the neutrino. The photon was talked about on the electromagnetic radiation section of chapter 1. Thomas Young discovered the wave theory in 1801, and the word *photon* might have been born in that time. However, history has Albert Einstein in 1905 proposing a quantum of light, which behaves like a particle and may have created the word *photon*. Recall that a photon has no mass and technically speaking, it is not a particle. But it is considered a subatomic particle from Einstein's explanation of particle-wave duality. And the second particle, the neutrino, was first suggested by Wolfgang Pauli in 1930. So we are starting here with five particles. Also, in 1931, Paul Dirac's mathematical equations led him to the concept of antiparticles with his finding of a positive electron—all particles have their own antiparticles. But let's stay with the standard particles for now. In the same year, mesons were discovered and started the meson theory of nuclear forces.

In 1946-1947, the word *lepton* was introduced to describe objects that do not interact too strongly (electrons and muons). In 1951, two new particles were discovered in cosmic rays and were named lamda^0 and K^0. In 1952, four particles were discovered with similar properties and were called delta^{++},

delta⁺, delta⁰, and delta. It was also the year that the bubble chamber was invented leading to an explosion in particle discoveries. After that, bosons (W^+ and W^-), and a second neutrino were discovered. In 1964, Gell-Mann and Zweig "tentatively" put forth the "idea" of "quarks" to satisfy mathematical projections. They suggested that mesons and baryons are composites of quarks. This was later proved to be partially right. By 1973, quarks were determined to be real particles. Also, gluons were introduced as massless quanta of the strong interaction field. In 1974, a J/psi particle was discovered. In 1976, a tau lepton was discovered. In 1989, experiments carried out in high-speed accelerators suggested that there are three and only three generations of fundamental particles. We will explore the meaning of this. In 1995, a sixth quark was discovered. Nothing is listed after 1995, but for sure, high-speed accelerator experiments continue. With the reality of quarks, it is now known that the meson particle is made up of two quarks and the group has as of now expanded to about 140 different particles. Also, groups called baryons, of which the proton and the neutron belong, are made up of three quarks, which now claim about 120 particles. So obviously, the subatomic family has zoomed well past the claim on the previous slide of about 100 particles.

Sorting out the Subatomic Particles

For the subject matter on the next series of pages, I am using several Web sites plus a magnificent wall reference chart that I purchased a number of years ago. It is titled the "Standard Model of Fundamental Particles and Interactions. An updated version is accessible on Web site *http://www. CPEPweb.org/*.

In the theme of simplifying this story I warn that much will be left out. If your interest level becomes tweaked, I refer you to some of the recent publications listed on the Reference page at the end of this book.

Quarks

It has been mentioned in chapter 1 that a proton consists of two types of quarks. Let's make this our starting point. There are actually a total of six quarks. They are found as three groups of two as shown on table 5. Only two of the six, the up quark and the down quark are found in the proton and the neutron. Quarks have mass and are electrically charged, but unlike all other particles, the quarks have fractional charges, the up quark being +2/3 and the down quark being -1/3. So the proton has two up quarks (UU) and one down quark (D) for a +1 charge, and the neutron has two down quarks (DD) and one up quark (U) for a net 0 charge.

The traditional teaching for the neutron is that it can be thought of as a proton and an electron. But now that we know that the neutron consists of three quarks that result in a zero charge, how do we conceptualize the

Flavour	Mass GeV/c^2	Electric Charge
up (u)	0.003	+2/3
down (d)	0.06	-1/3
charm (c)	1.3	+2/3
strange (s)	0.1	-1/3
top (t)	175	+2/3
bottom (b)	4.3	-1/3

Table 5. Six Quarks—Six Fundamental Particles

electron? The answer requires that instead of thinking of the neutron as a combination of a proton and an electron, we need to think of the neutron as not having an electron. We need to think of the neutron as two down quarks and one up quark (DDU). Its charge equals zero. It is happy with a zero charge. However, if the neutron goes through beta decay (recall radioactivity, chapter 1), the neutron turns into a proton by having one down quark turn into one up quark (DDU), as shown in Illustration 6. This is common with isotopes, which may have an unstable nucleus due to either an excess or a deficiency in neutrons. In this process of transforming one quark into another, mass is lost. As a result, an electron, an antineutrino, and energy are emitted. This transformation through beta decay is mediated by the boson W⁻. In addition to the W⁻, the W⁺ and the Z⁰ act as exchange particles. In doing so the charge turns from zero to a +1. So visually, the decay products are one proton and one electron—and there emerges that sneaky little electron.

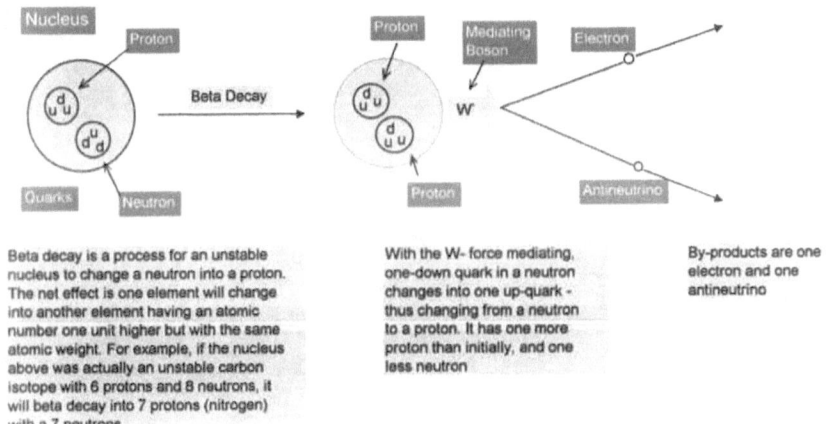

Beta decay is a process for an unstable nucleus to change a neutron into a proton. The net effect is one element will change into another element having an atomic number one unit higher but with the same atomic weight. For example, if the nucleus above was actually an unstable carbon isotope with 6 protons and 8 neutrons, it will beta decay into 7 protons (nitrogen) with a 7 neutrons.

With the W- force mediating, one-down quark in a neutron changes into one up-quark - thus changing from a neutron to a proton. It has one more proton than initially, and one less neutron

By-products are one electron and one antineutrino

Illustration 6. Neutron Beta Decay

Continuing with quarks, refer again to the quark table 5. Four more quarks were discovered later on, the last two as late as 1994. From the previous slide, the up quarks and the down quarks make up the protons and the neutrons and are the least massive of the six quarks. They are the only ones to occur naturally. Continuing in order of increasing mass are particles charm and strange, and top and bottom. They work in pairs. The names have no significance. They also have charges of either +2/3 or -1/3 and will be found in groups of 2 (mesons) or 3 (baryons) in a way that results in a unit charge of +1 or -1. As mentioned before, the proton and the neutron belong to the baryon class since they consist of three quarks. There are about 120 others. Mesons consist of two quarks and together amount to about 140 different mesons. Since they consist of quark arrangements, none of these particles are elemental.

Leptons

Refer to table 6. Six particles exist: the electron and electron-neutrino, the muon and the muon-neutrino and the tau and the tau-neutrino. The three neutrinos differ in mass but are similar in that they carry no charge. The electron, muon, and tau carry a negative charge. The muon and tau are much heavier than the electron and do not exist in everyday matter because they decay very quickly. Unlike quarks, which exist only in groups of two or three, the leptons exist as single particles. All six particles have antiparticles for a total of twelve leptons. They are all subject to the weak interaction. The charged electron, muon and tau particles are also subject to the electromagnetic force. Scientists don't expect to find anymore. These particles are elemental.

	Mass Ge/c^2	Electric Charge
Electron Neutrino (v_e)	$<1 \times 10^{-8}$	0
Electron (e)	5.1×10^{-4}	-1
Muon Neutrino (v_u)	$<2 \times 10^{-4}$	0
Muon (u)	0.106	-1
Tao Neutrino (v_t)	<2.02	0
Tao (t)	1.7771	-1

Table 6. Six Leptons—The Six Other Fundamental Particles

Exchange Particles

Then there are the exchange particles shown in table 7 referred to as force carrying, interactive, intermediating, etc., which deal with the four forces: gravity, electromagnetic, weak, and strong, making them all the forces in the universe. They are the following:

1. The pions are particles with mass and are responsible for the strong force. There are three pions each belonging to the meson group. The three exchange pions have electric charges of -1, 0, and +1.
2. The photons, which we have already discussed, are also massless particles and are responsible for all electromagnetic radiation forces.
3. The bosons are actually of considerable mass and consist of three particles (two W particles with charges of -1 and +1, plus a Z^0 particle with no charge). They are responsible for the weak forces within the atom and for neutron beta decay.
4. The graviton is a massless particle and is yet to be discovered. It is theorized to be responsible for all of the gravitational forces.

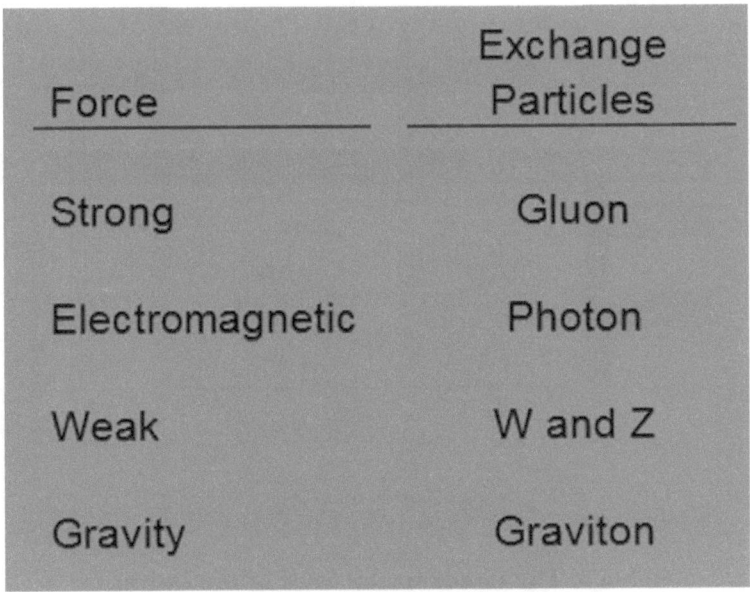

Force	Exchange Particles
Strong	Gluon
Electromagnetic	Photon
Weak	W and Z
Gravity	Graviton

Table 7. The Force Carrying Particles

Summary

The 1930s was a period of major accomplishment in the world of science. For over 2,000 years, philosophers and scientists struggled to understand the smallest fragment of matter. This fragment was referred to as *atomis* or the *atom*, way before there was any comprehension of its content. But in the 1930s, the secrets were unveiled. The nucleus consisted of protons and neutrons and was encircled by electrons. The protons and electrons were held together by attracting electrical charges. They were considered fundamental at the time, meaning incapable of further reduction. But then, quarks were discovered and then antiquarks and then leptons and then force-exchange particles, and with it a new set of rules, a new semblance of order.

This new order is combined on table 8. Are we finally there—are quarks, leptons and force particles really elemental? There appears to be some consensus that it is so. But truth be told, it is only known that these new particles are less than 10^{-19} m in radius. In Chapter Six we will explore an exciting theory about superstrings. They will challenge our thinking about the smallest of the small.

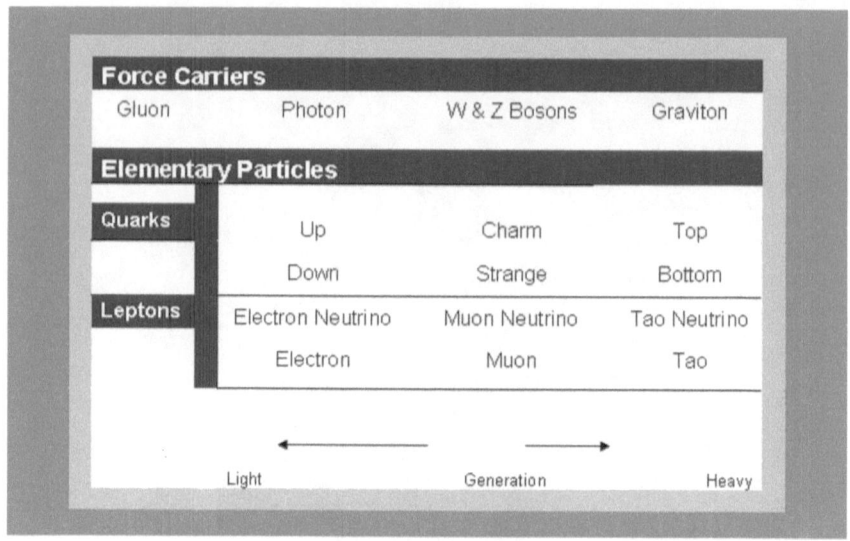

Force Carriers			
Gluon	Photon	W & Z Bosons	Graviton

Elementary Particles			
Quarks	Up	Charm	Top
	Down	Strange	Bottom
Leptons	Electron Neutrino	Muon Neutrino	Tao Neutrino
	Electron	Muon	Tao

Light ← Generation → Heavy

Table 8. The Fundamental Stuff of the Universe

Particle Accelerators

The question "how were most of the particles in this chapter found?" is certainly a valid one. The only way that scientists can study the inner workings of the atom is to take them apart. Since atoms are bound together very strongly (remember the strong force as one of the four fundamental forces?), it's easier said than done. But scientists are very clever. They started smashing atomic nuclei into other atomic nuclei just to see what happens. It was probably a little more scientific than that. They started by isolating, let's say, a hydrogen nucleus, which is simply a proton and perhaps a neutron or two (in the case of deuterium and tritium). Since it has a positive charge, it is sent into a positively charged chamber and through the use of repulsion, accelerated the proton to extremely high speeds. The devise can be linear or circular. When properly accelerated, the proton is aimed at another proton, and the fragments are examined. Now since the fragments can't be seen, we will have to research how these clever scientists *observed* them. The very first accelerator was built in 1929 at the University of California, Berkeley. It was a 9-inch cyclotron, and it accelerated a proton with an energy level dwarfed by today's standards.

Over the years, particle accelerators have gotten bigger, faster and more powerful. One is the Stanford Linear Accelerator in California (SLAC), which is 1.8 miles long. Another is the Fermi National Accelerator (Fermilab) circular accelerator in Illinois, which covers 10 sq miles.

All particle accelerators have the following crucial parts:

- particle source
- copper tube—within which is a vacuum and the particle electromagnets—keep the particles to a narrow beam
- targets
- detectors
- vacuum systems
- cooling systems
- computer/electronic systems—to analyze the data
- shielding—protects workers from high radiation levels
- monitoring systems—for safety purposes
- electrical power system

Particle Sources

A particle source must be an electrically charged particle so that it can be accelerated by changing the electrical source. Protons, electron, positrons (positive electrons, also called antimatter), ions, and nuclei of heavy metals, such as gold are examples of some of the sources. An electron gun uses a laser to knock electrons off of a semiconductor. The electrons then enter the copper accelerator tube, which is under vacuum.

Targets

As the electron beam travels through the copper tube, it is focused by the use of magnets into a concentrated stream. Targets vary with the desired experiment and range from metal foils to other head-on streams of particles.

Detectors

The detectors are the main event. They give purpose to the experiment. They are the tricks that allow the experimenter to see these otherwise invisible-to-the-eye particles. One such detector is a bubble chamber, which is a liquid gas in its chamber. As particles go through the chamber, they vaporize the liquid leaving a bubble trail. Photographs leave a telltale fingerprint. Another detector is a cloud chamber, which is similar in concept, but the vapor is ionized by a passing particle. And thirdly, there are solid state electronic detectors. The detector at Stanford has a detector that is more than six stories high and weighs over 4,000 tons.

When the accelerators came on the scene in the 1950s and 1960s, scientists were able to go way beyond the initial three elemental particles, the proton, the neutron and the electron to hundreds of new particles that were even smaller. As the accelerators got more powerful with higher energy beams, even more particles were found. Most of these particles only exist for a billionth of a second—but got caught by the bubble chamber cameras.

As bigger and better particle colliders are being built, scientists go back further in time to what might have happened near the beginning of the big bang. There have been many winners of the Nobel Prize in Physics award as a result of work done on these particle accelerators. I list them here from http://nobelprize.org/nobel_prizes/physics/laurates/1959/index.html.

- 1959—Emillio Segre and Owen Chamberlain for the discovery of the antiproton
- 1976—Burton Richter and Samuel Chao Chung Ting for the discovery of a heavy element particle of a new kind
- 1984—Carlo Rubbia and Simon van der Meer for their contributions leading to the discovery of the field particles, W and Z, communicators of the weak interaction
- 1990—Jerome Friedman, Henry Kendall, and Richard Taylor for their work done, which have been of essential importance for the development of the quark model in particle physics
- 1995—Martin Perl and Frederick Reines for their experimental contributions to lepton physics.

The most exciting piece of news in the particle accelerator community is the recent completion of the Large Hadron Collider (LHC) at CERN, located on the border of Switzerland and France near Geneva. It is the world's largest and highest energy supercollider and will accelerate two opposing beams of protons to 99.999999% the speed of light. It is 16.7 mi long and resides about 300 ft below ground level (refer to *http://en.wikipedia.org/wiki/Large_Hadron_Collider*). Scientists hope it will shed light on fundamental questions in physics. Dr. Tara Shears of the University of Liverpool was quoted as saying on BBC News that "We will be looking at what the universe was made of billionths of a second after the Big Bang." (This was posted at *http://news.bbc.co.uk/2/hi/science/nature/7604293.stm*). The project was funded and built in collaboration with over 10000 scientists and engineers from over100 countries as well as hundreds of laboratories.

Its first day of operation was actually September 10, 2008 and proton beams were circulated in the main ring for the first time. However, on the next day a serious fault relating to a helium leak between two superconducting magnets shut the LHC down. It will start up again in the spring of 2009.

There are approximately 70 particle accelerators in the world today. Europe leads in this department with 33. Of the Europeans, Germany leads with 14, and France is second with 3. Ten other European countries have 1 or 2 accelerators. The United States has 20, and Canada and Mexico have 2. Asia has 11.

Chapter Three

Important Fundamentals in Physics

The Four Forces in Nature
Einstein's Mass/Energy Equation
Quantum Mechanics
Einstein's Relativity

1. The Four Forces in Nature

THERE ARE FOUR fundamental forces in the universe, no more, no less. Atoms, planets, stars, galaxies, atomic fission, neutron stars, black holes, the big bang, humans, and everything else are all governed by these forces. In Einstein's days, the strong and the weak forces had not yet been discovered. With the two that had been discovered, gravity and electromagnetism, Einstein set out on a thirty-year commitment to unify them. He was unsuccessful, but his dream was carried on more than half century later when scientists considered the unification of the four forces the holy grail of modern physics. Today all, except gravity, have been unified.

Three of the forces deal directly within the atom: the strong, the weak, and the electromagnetic. These are the three forces that have been mathematically unified today. Gravity, however, still eludes and works mostly on the scale of the cosmos. Gravity is actually so weak that it can be discarded as a factor in the world of subatomic physics. Yet it is strong enough to affect us in everyday life, not the least of which is holding us onto the surface of planet Earth. In order to escape earth's gravity, according to Wikipedia, a rocket needs to achieve a speed of 6.96 mi/sec. That's roughly 25000 mi/hr, more than 10 times the speed of a rifle bullet (Note that this conflicts with *Brave New Universe*, Paul Halpern and Paul Wesson, p. 68 which claims a speed of 17000 mi/hr).

Without gravity, the planets and stars would never have emerged in the first place, and life would not have evolved. Electromagnetism also affects us on a daily basis from visible light to microwave ovens to radios and TV and cell phone transmissions, to infrared rays (heat) to ultraviolet rays (sunburn) to x-rays (dentist) and so on. On the other hand, the weak and strong forces have no effect on our everyday life. So let's proceed to expanding on these forces.

Gravity: ($F = Gxmm'/d^2$)

In1687, Isaac Newton proposed his universal law of gravity. We've all heard the tale of the apple falling from the tree and inspiring Newton's genius. Gravity acts upon all particles that have mass. It might seem contradictory that gravity acts on all matter while the graviton, the facilitator of the force is massless. Nevertheless, that is exactly the situation. The range of gravity is infinite—you, your mass, actually have a gravitational effect on all of matter including the faraway galaxies. But your effect would be infinitely small. Newton's equation says that the gravitational force between two objects (F) is equal to the gravitational constant (G) times the product of mass 1 and mass 2 (m and m') divided by the square of the distance between them (d^2), the inverse-square law. All you need to know is mass and distance. Noteworthy is that the gravitational constant is an exceedingly small number at 6.67×10^{-8} $cm^3/gm\text{-}sec^2$. This means that unless the masses are exceedingly large and/or the distance between the two bodies are relatively close, the resulting force will be exceedingly small.

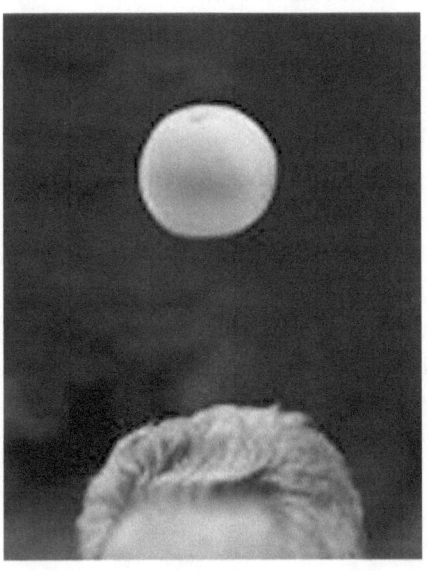

Here is an example. Let's calculate the attraction of two balls positioned on a smooth, perfectly level surface, 10 inches away from each other. Let's say that the balls have a mass of 10,000 g each (slightly heavier than a bowling ball).

So, from Newton's equation:
$$F = G (10000 \text{ g}) (10000 \text{ g})/(25.4 \text{ cm})^2 = 155038 \text{ gm}^2/\text{cm}^2.$$

This seems like a big number until we factor in that tiny gravitational constant (G) to get a gravitational force F of only 0.01033 dynes. This is an extremely small number as forces go and explains why we don't see objects flying around the kitchen or anywhere else on earth.

Let's calculate the actual gravitational force between earth and our moon. Earth weighs 5.98×10^{27} g, and the moon weighs 7.348×10^{25} g. They are separated center-to-center by 3.844×10^{11} cm. Fill in the blanks and factor in G and you get an incredible 20×10^{24} dynes. You can see that the force values get quite staggering once we get to factoring in large objects. This gravity force value is sufficient to keep the moon in its orbit.

Just one comment before we get on to the ultimate question on gravity. While gravity may seem very strong in order for it to facilitate the birth of the stars and planets and to keep them in orbits within billions and billions of galaxies, it is really quite small relative to the electromagnetic force. How small is it? This is an example. The author of this story is John Walker. He put a small cubical magnet in the jaws of a pair of small pliers. The magnet was 4 mm³ (about 0.16 in³). It is small! He located a spherical steel ball that weighed 550 g (just over 1 lb) and proceeded to pick the ball off the ground with the little magnet. Consider what's being done here. A 1-gram magnet pulled an object weighing over 1-pound off the ground with the electromagnetic force while an opposing gravitational force—the entire planet earth—was pulling downward with a weight of 6×10^{27} grams and a volume of 10^{27} cm³. The winner—electromagnetism! (*http://www.fourmilabch/gravitation/foobar/*).

So now we get to the million-dollar question, "*Since electrical attraction is not involved in the force of gravity, what is it that attracts one chunk of matter with every other chunk of matter in the universe?*" How do we conceptualize this? Is there some kind of subatomic string attached to every atom? Are these questions strictly philosophical, or can science rescue the day?

In describing gravity, scientists liken it to electromagnetic radiation, which is often presented as an electromagnetic field. The latter is easy to demonstrate visually as all of you have seen somewhere in your educational background, by spreading iron shavings around a magnet. It is quite a dramatic demonstration because how else would you know if the force field existed or what the force field would look like. If you could do that on a grander scale, say dump iron shavings around earth, you would see the same kind of force field. And, in fact, the entire universe is awash in an electromagnetic field. Well, think the same with gravity—it is a force field without an electrical charge. Unfortunately, we have no way to see the gravitational field. The field is more mechanical, if you will, with massless gravitons enveloping like a fabric around us, pulling us down every time that we jump up, and keeping all the universes, planets, and stars in their orbits and pulling on us from every piece of matter in the universe. At least this is the latest thinking. Remember, gravitons have yet to be discovered.

Permit me the audacity to try to explain the genius of Einstein on gravity. Just kidding—actually, I will be leaning quite heavily on Brian Greene's, *The Fabric of the Cosmos: Space, Time and the Texture of Reality*, cited on the References page.

In 1916, Einstein published his general theory of relativity, which followed his special theory of relativity in 1906. It basically deals with motions caused by the attraction of all kinds of matter and energy. Regarding gravity, it describes how materials produce and respond to changes in space-time geometry to the curvature of space-time by matter. To this day, Einstein's general theory is still considered, according to Brian Greene, the most accurate and elegant description on the workings on gravity, 102 years later.

So what is all this space-time-warp-fabric stuff all about? The "fabric" I take to mean the universal network of gravitons. Warps only have meaning in the presence of mass and energy. Matter makes the fabric bend (warp) much like a bowling ball on a mattress. In space, one body approaching another, say an earth-like body approaching a sun-like body, at some just right distance would get caught up in the warped space fabric and burn up as it nears the sun. As Greene explains, "It's as if matter and energy imprint a network of chutes and valleys along, which objects are guided by the invisible hand of spacetime fabric." Einstein spent many years defining mathematically the precise shape and size of the warping as a function of the quantity of matter and energy. They are called the Einstein field equations. These equations also tell us that gravity does not travel from body to body instantaneously, but precisely at the speed of light. If our moon mysteriously disappeared, the gravitational effects on earth would not be experienced for 1.5 seconds late. Likewise with the sun, except the effects would be 8 minutes later.

Electromagnetism

Some of this was covered in chapter 1. I will recap somewhat. First, electromagnetism, like gravity, is a force that affects us on a daily basis. But let's go inside the atom for a bit. The electrical force between the nucleus and the electron(s) is just sufficient to keep the orbiting electrons from flying off. I say *just sufficient* because some of the outer shell electrons can be easily rubbed off. If energy is applied to an atom in the form of heat (e.g., infrared radiation photons), electrons circulate faster causing them to jump to a higher energy level. Electrons in the outer orbit may actually jump off. Envision that while some electrons jump to a higher level, others will be cooling and returning to a lower energy level. In the process of returning to the lower level, a photon is given off—in the form of electromagnetic radiation. We learned of the electromagnetic scale of which visible light is but a very small portion. The electromagnetic force then is the attraction of the electrons to the protons, and since electrons are negatively charged, they repel each other while in orbit but are held on to the atom by the attraction of the protons. The attractive and repulsive forces are also mathematically similar to the gravity equation (the inverse-square law) in that their forces gets weaker and weaker, the farther apart the particles are. Unlike gravity, electromagnetism also has a repulsive force (save this thought for chapter 4).

The carrier force (particle) for electromagnetism is the massless photon—much like the massless graviton in gravity. Both electromagnetism and gravity have infinite range. The electromagnetic force, however, is much, much stronger than gravity—by a factor of 10^{36} on a relative scale.

Weak

This force is actually an interaction, which is facilitated by the three bosons, two Ws and one Z. If this sounds familiar to you, that's good because a classic example of the weak force is beta decay, which we already talked about in chapter 1. Recall that a neutron is converted into a proton by converting one down quark into one up quark facilitated by a boson. In the process, the boson is broken down into one electron and one antineutrino. The mean life of a boson is about 3×10^{-27} sec, which means that the range of the weak force is only 10^{-18} m—about 1,000 times smaller than the size of the atomic nucleus. Since the weak interaction is both very weak and very short range, its most noticeable feature is the changing of the neutron into a proton through the quark change from down to up. There are actually two other boson type interactions, but we need not go into them.

Strong

The strong force is also called the strong interaction. This force is facilitated by gluons acting upon quarks, antiquarks, and gluons themselves in a way that defeats the repulsive forces of proton to proton. In so doing, it ensures that the nucleus does not explode in what is otherwise a very hostile environment of positive charges. First, the strong force is the strongest of the four forces being 100 times that of the electromagnetic, 3,000 times that of the weak and 10^{36} times that of gravity. It also works in the shortest of distances, that being about the size of the nucleus itself. Refer to table 9.

Relative Strengths of the Four Forces	
Strong	1
Electromagnetic	0.01
Weak	0.003
Gravity	10^{-36}

Table 9. The Four Forces of Nature

2. Einstein's Mass/Energy Equation: ($E = mc^2$)

Einstein's mass/energy equation is considered the most famous equation of all time and is as close as any equation of physics can be to having global household recognition. Yet how many of us really understand it? Not many probably, unless you are part of the scientific community. It's actually rather simple.

The simplicity of the equation is *energy equals mass times the velocity of light squared*. It is as profoundly important as it is elegantly simple. Energy and mass are tied together. The equation can be expressed as $E/m=c^2$. The speed of light is a constant. It never varies. It is also a big number, 300,000,000 m/sec. Seven and one half times around earth in one second! Now, for the equation the speed of light must be squared. So, $c^2 = 9 \times 10^{16}$ joules/kg.

Calculations:

$E = mc^2$

$E = m \, (9 \times 10^{16})$

$E/m = (9 \times 10^{16})$

It also tells us that a very small mass results in an extremely high energy value, and that if you know the mass of a body you can easily determine its energy. For example, one gram of mass is equivalent to the following amounts of energy:

82 billion BTUs
215,000 tons of TNT
21.5 tons of kilocalories.

(http://en.wikipedia.org/wiki/Mass-energy_equivalence)

3. Quantum Mechanics

Max Planck put forth his quantum theory in 1900. It was further developed throughout the first half of the twentieth century as a mathematical/theoretical tool to better explain the workings of the atom. Physicists involved in the development of quantum mechanics reads like a science hall of fame—Albert Einstein (relativity), Werner Heisenberg (uncertainty principle), Max Planck (quantum hypothesis), Niels Bohr, Erwin Schröedinger (equation), Max Born, Paul Dirac, Wolfgang Pauli (exclusion principle), and others. Schröedinger is credited with developing the first equation in 1926 upon which quantum mechanics is based. Prior to quantum mechanics, Newton's laws of motion and Maxwell's laws of electromagnetism predicted, against reality, that electrons should not stay in their orbits. Their predictions were that the electrons should quickly move toward and crash into the nucleus. In other words, the classical laws at the time were not working, at least on the atomic level. Quantum mechanics was developed to deal with the unfamiliar features of the atom and subatomic particles. They became suitable for such terms as uncertainty, quantum fluctuations, wave-particle duality, vector space, probability, and quantum entanglement.

If this is clear as mud, don't feel bad. "By 1928 or so, many of the mathematical formulas and rules of quantum mechanics had been in place and, ever since, it has been used to make the most precise and successful numerical predictions in the history of science. But in a real sense those who use quantum mechanics find themselves following rules and formulas laid down by the 'founding fathers' of the theory—calculational procedures that are straightforward to carry out—without really understanding why the procedures work or what they really mean. Unlike relativity, few if any people grasp quantum mechanics at a 'soulful' level". (*The Elegant Universe*, Brian Greene, p. 87).

Greene then advises, "If along the way (to his next subject), quantum mechanics seems to you to be altogether bizarre or even ludicrous, you should bear in mind two things. First, beyond the fact that it is a mathematically coherent theory, the only reason we believe in quantum mechanics is because it yields predictions that have been verified to outstanding accuracy. Second, you are not alone in having this reaction to quantum mechanics. It is a view held to a greater or lesser extent by some of the most revered physicists of all time. Einstein refused to accept quantum mechanics fully. And even Niels Bohr, one of the central pioneers of quantum theory and one of its strongest proponents, once remarked that if you do not get dizzy sometimes when you think about quantum mechanics, then you have not really understood it." (*The Elegant Universe*, Brian Greene, p. 88)

Today, quantum mathematics operates on the scale of the atom and smaller—much, much, smaller. On the other extreme are Einstein's equations of relativity, which operate on the scale of the cosmos. Therefore they clash and prevent the unification of the four forces. While the quantum field, when factored in with the gravitational field and time has been likened to anything out of control—wild, frenzy, orgy, turbulent, well, you get the picture. Anything larger than a microscopic picture is considered to be a smooth model of general relativity. What this means is that calculations that try to merge the equations of relativity with those of quantum mechanics run into answers of infinity.

4. Relativity

Albert Einstein, at the age of 26, completed his *special theory* of relativity and in doing so turned the world of space and time and motion into a new way of conceptualizing the universe. At its core were properties of light and relative motion as defined at constant velocity. *Special relativity* does not address gravity and therefore is a suitable mathematical model whenever gravity can be neglected. Later in 1915, Einstein completed general relativity dealing with the workings of gravity and acceleration.

I must warn you that what you will read on the next few pages on this subject will challenge what we on earth call common sense and will give you a feeling of smoke and mirrors, and sleight of hand. Don't feel bad. You have a lot of company. It appears that common sense does not work well on cosmic levels. Think of this as a tribute to the genius of Einstein. He is often credited not only as a great mathematician and philosopher but also as a free thinker with imagination and intuition in his approach to science. This is where he excelled. The concepts attributed to his creative thinking have withstood the test of time with over one hundred years of scientific scrutiny and confirmation from his peers. So soak it up.

If you remember nothing else, memorize the following six rules as fundamentals of relativity: (1) the speed of light is constant; (2) motion in space is relative; (3) the fabric of space is warped, and light traveling through the warp will bend; (4) at very high speeds, mass will increase; length will shorten and time will slow; (5) gravity and acceleration are intertwined; and (6) the universe has to be changing—either expanding or contracting.

Albert Einstein was born in Germany in 1879. As a child, he had a natural inquisitiveness about the workings of nature. "Albert at the age of twelve was presented with a book on Euclidean geometry. Albert marveled at the crisp certainty of the mathematical arguments presented. Soon he learned how to construct his own proofs, creating geometric rules from simple propositions." At the age of 16, he imagined himself running after a beam of light, faster and faster and faster until he caught up. He wondered if the beam would appear stationary. Newtonian physics at the time would say yes, the light beam would appear stationary. However, this contradicted the results of Maxwell's equations. Einstein's special relativity delivered the answer with his equations that preserved the *constancy of the speed of light* by asserting that measured distances and durations depend on the relative velocities of the observer and the observed. Hang in there. (*Brave New Universe*, Paul Halpern and Paul Wesson, p.37)

This example is from Brian Green's *The Elegant Universe*. I have shortened this story for the sake of simplicity, and changed the names of the featured stars to my grand-children's names. Here is how constancy works. Suppose you, Antonia, put on your fast-running sneakers and wait for nighttime to turn on a large spotlight. Your brother Joey is with you. You both have stopwatches. When the spotlight is turned on, the light beam will zoom instantly to 186,000 mi/sec (7½ times around earth in a second). You are already 7½ earth laps behind in one second. So with great haste you say goodbye to your brother and begin running after the beam—faster and faster and faster. You get to 80% of the speed of light and maintain velocity. You don't know or feel this, but your metabolism and your stopwatch have slowed relative to your brother, the observer. You would think that the light beam would only be going 20% faster than you because that would seem logical. But to your surprise, the beam is still going past you at 100%—the full speed of light! You don't know that your time has slowed unless you compared your watch with your brother's, whose watch as an observer has not slowed. When you realize that you lost a substantial amount of time you are, in effect, traveling slower—in fact 80% slower. Therefore, the light beam relative to you appears to be and is traveling at 100% the speed of light, even though you are traveling at 80% the speed of light. I warned you! This is constancy! Regardless of relative motion between the source of photons and the observer, the speed of light is always the same. No matter how hard or fast you want to chase a light beam, it still retreats from you at the speed of light. The key to understanding this lies in the fact that time is slowed significantly at high speeds—this is not an illusion, but it is a true paradox. I am sure that if all of us were advanced mathematicians looking at the purity of Einstein's equations, we would have a greater appreciation for this concept. (*The Elegant Universe*, Brian Greene, p. 32)

Time, distance, and mass are all necessary to describe the mechanics of the physical universe.

First, let's define mass and weight. Both reflect the accumulation of all the atoms making up an object. However, weight is specific to the force of gravity, mass is not. Weight for us on earth is the force of gravity on us. We need to specify weight as earth weight because we will weigh less on the moon, and on a rocket in space we will weigh zero.

Mass is described as the resistance to a change in motion or the force required to cause a body to accelerate. It is independent of gravity, meaning that an object will have the same mass whether it is on earth, on the moon or in space. Now having said that, I hope you are saying it is not so because in Einstein's time-space continuum, we have seen that an object in motion will increase in mass directly with an increase in speed. While this does not become significant until very high speeds are achieved, relative to the speed of light, the statement still stands. At the speed of light, the mass of an object would, mathematically speaking, approach infinity.

In addition to mass, distance and time are also warped at very high speeds. The length of on object, such as a spaceship, would decrease with speed and approach zero at the speed of light. And time would likewise go slower.

The effects then of high-speed motion on matter are called dilations—slower time, heavier mass, and shorter length. Don't worry. There is nothing

on earth that moves that fast (except light) and although the above effects are real on earth, they are negligible. Let's go through an earthly example put forth again by Brian Greene. Again this is a shortened version. Joey and his sister Antonia went to the racetrack to test his new race car. Joey speeds down the one-mile track at 120 mi/hr while Antonia on the sideline times him. Joey too has a stopwatch. Prior to Einstein, no one would have questioned that both stopwatches would record exactly the same time. But according to special relativity equations, while Antonia measured 30.00000000000000 seconds, Joey measured 29.99999999999952 seconds. Now obviously not even the best of clocks could measure to that accuracy, but it serves the purpose of this example. It is a mathematical exercise. Likewise, Joey measured his car at 16 ft, but Antonia by calculating from his speed and time would get an answer of 15.99999999999974 ft long. Keep in mind that if Joey could somehow increase his speed to 580 million mi/hr, 87% the speed of light, the mathematics of special relativity predict that Joey's car would measure 8 ft in length. Similarly, the time to traverse the drag strip would be about twice as long as the time measured by Antonia's stopwatch. (*The Elegant Universe*, Brian Greene, pp. 26-27)

In the entire universe, there is no absolute reference point to observe from, meaning simply that all matter is moving. Everything in the universe is expanding, galaxies revolve and planets revolve and rotate. Earth itself rotates on its axis every 24 hours, revolves around the sun once every

365½ days at an average speed of 66,000 mi/sec, and our solar system is screaming around its galactic center. Nothing is stationary! This means that for any event in the universe, observers from different viewpoints (different planets) will interpret the event differently. Here is a story that I remember reading about back in my youth. A lady is seated in the passenger car of a moving train. An object falls from above within the railcar and hits her on the head. It causes damage, resulting in a court hearing. Testimony by four witnesses, tell four different stories. A fellow passenger alongside the plaintiff said that the object fell down in a straight line. A witness who was observing the train from a stopped vehicle described the path of the falling object as curved. A third witness from Mars said no, the path of the object was clearly elliptical. And lastly, a witness from our nearest galaxy saw the path of the object as para-hyperbolic (or something like that). And all of them were correct, relatively speaking. *There is no such thing as an absolute frame of reference in our universe.*

Let us go back 250 years and reexamine Newton's first law, which says that *a* body in motion will remain in motion until acted upon by outside force. This is before Einstein's dilation effects were discovered. What a statement though—until acted upon by an outside force! Everything on Earth, of course, is affected by the forces of gravity, air, wind, etc. So for Newton to conceptualize this law as in a space environment was a stroke of pure genius. So what this means is, if I go far enough out in space, away from any gravitational effects, wind effects or any other forces and throw a baseball, it would literally travel forever until acted upon by an outside force.

Einstein's special relativity concerns itself with all of the physical laws in nature. It starts with *motion is relative.* We are talking about constant velocity and force-free motion and have meaning only in comparison with other objects or individuals also undergoing force-free motion. First, we cannot feel or know that we are in constant velocity motion without a reference landmark. In a car, you know that you are moving because you see reference points, trees, buildings, etc., as stationary objects as you go by. You also have the feel of the road and with most cars the noise of the engine. So you are not fooled. But even with a car, most of us have experienced relativity on probably more than one occasion. I'm in a car on a four-lane road, two lanes in each direction. I stop at a red light. A car in the lane next to me approaches in my direction and stops alongside of me. I am distracted and in deep thought when I look over to see that my car is moving forward. In panic, I stomp on the brakes, before I realize that I already had my foot on the brake pedal. I panic again in a state of confusion—only to realize that I was totally stationary and the car next to me was moving in reverse. This all happened in seconds, but it seemed so real that I was moving forward.

On airplanes, we experience acceleration during takeoff and during the climb to 33,000 ft, where the pilot levels the plane and completes acceleration to a velocity of 550 mi/hr where he announces that we can now safely move about the cabin. During acceleration, we are slammed to the back of our seats as well as to the bottom of our seats due to gravity. Once the plane achieves constant velocity, we are only subject to gravity (theoretically less so then on the ground, but not enough to make much difference), but we are no longer slammed to the back of our seats—acceleration is now zero. Now if you close your eyes and if the flight is exceptionally smooth and if the engines are somewhat quiet, you can almost convince yourself that you are not moving. You look out the window and see that the ground is moving very slowly, or is it the plane that is moving? This is Einstein's playground. There is no way of knowing the state of motion without some direct or indirect reference points.

With special relativity, Einstein proved the constancy of light and defined new rules for objects in motion at constant-speed in space-time. But the constancy of light proved to be incompatible with Newton's universal theory of gravity from the later 1700s. After ten years of dedicated effort, Einstein put forth general relativity in 1915 to address this incompatibility and address other issues arising from special relativity.

Newton was born in 1642 in England and was a giant of science in his time. Of his many achievements, we will focus on his workings in gravity. Newton is credited for identifying gravity as the single force responsible for all of the mechanics of matter in the universe. He declared that all of matter is subject to the invisible gravitational force. He alone asserted that the attraction between two objects is a function of the masses and the distances between them. Specifically, the force is proportional to the product of their masses and inversely proportional to the square of the distance between them. Implicit in this equation is a universal constant. The agreement over the years of calculated predictions and the observations has been remarkable.

So what is this incompatibility between Einstein and Newton? It starts here—nothing travels faster than light! No other signal can get to us faster—OK? Yet Newton's equation of gravity does not deal with a time dependency. So if one object moves relative to another, then according to Newton, the mutual gravitational force due to the change in motion must be assumed to happen immediately! Carrying this logic out to its illogical conclusion, if the sun blew up right now, we would feel the tremendous earth changes in gravity right away according to Newton while we on earth, in reality, would not see the change for 8 minutes, the time for light to get to earth from 93 million miles away—an obvious incompatibility that even we can easily understand.

So though Newton's equations have been tremendously successful for over 200 years, Einstein was sure of his special relativity and went ahead, committed to disprove Newton. We know that Einstein, the best mind of our times, spent ten years committed to this project. So obviously, this was not an easy task. Let's see if we can understand this.

Einstein had to dispute Newton by constructing a new theory of gravity compatible with special relativity. Although Newton's equations were a huge success, even he had doubts about how gravity works. In his words:

"It is inconceivable, that inanimate brute matter, should, without the mediation of something else, which is not material, operate upon and effect other matter without mutual contact. That Gravity should be innate, inherent and essential to matter so that one body may act upon another at a distance thro' a vacuum without the mediation of anything else, by and through which their action and force may be conveyed, from one to another, is to me so great an absurdity that I believe no Man who has in philosophical matters a competent faculty of thinking can ever fall into it. Gravity must be caused by an agent acting constantly according to certain laws, but whether this agent be material or immaterial, I have left to the considerations of my readers!" (*The Elegant Universe*, Brian Greene, p. 57)

In 1907, Einstein had another "insight" that would eventually lead him to a radically new theory of gravity. It would later be called the equivalence principle, which recognizes accelerated motion and gravity as interwoven.

Suppose a cubicle is traveling at a constant rate of speed in space. With no acceleration, no gravity, and no benchmarks, the space traveler, as you know, cannot tell how fast the cubicle is moving, or in fact, if the cubicle is moving at all. Suppose further that the cubicle was programmed to initiate jet propulsion on the outside "bottom" (not a good word for there is no up or down in space) of the cubicle, unbeknownst to the space traveler.

The jets are set to accelerate at 32 feet per second per second (precisely the falling rate of gravity on earth). The pilot is standing when this happens and immediately recognizes the force on his feet and correlates it to that of earth's gravity. In fact, he feels that he just made a very soft landing on earth. He would have no way of knowing what really happened. He could not distinguish acceleration from gravity! *This is equivalence.*

Why did this realization please Einstein so much? Recall, gravity alone eluded unification of the four forces and in this regard is mysterious and an overwhelming challenge. It permeates the entire universe. It knows no boundaries. Now Einstein had correlated it to acceleration, which is much less complicated. With this fundamental understanding, Einstein knew that he could now begin to develop his new conceptualization of gravity.

Brian Greene comments, "Einstein, wrestling with this subject, wrote to a fellow physicist, 'I am now working exclusively on the gravity problem . . . one thing is certain—that never in my life have I tormented myself anything like this . . . Compared to this problem, the original (i.e. special) relativity theory is child's play." (*The Elegant Universe*, Brian Greene, p. 62)

How to simplify? The next breakthrough in his thinking was in 1912, and it focused on a particular example of accelerated motion. This involved a change of motion in a circle but at constant speed. I didn't bring this up before, but *acceleration* is defined as either a change in speed or a change in direction. We understand speed acceleration as forces that we can feel, even inside of a cubicle. Now let's go to another analogy by Brian Greene in *The Elegant Universe*. Now go to one of your favorite rides at any summer fair. Let's call it a zoom-wheel: it is horizontal to the ground and is shaped as a circle with a vertical wall against which riders Antonia and Joey stand. As it begins to rotate slowly Antonia and Joey begin to feel pushed ever against the wall and then more so as it builds speed. The force is such that the floor can safely drop out from underneath their feet because centrifugal force has become greater than gravitational force. If you could take this ride to outer space and spin it at the force of 32 feet per second per second it would mimic earth. And, in fact, that is what some of our space stations do.

I am going to apologize here before we start because the basic premise here will seem to contradict with what we studied previously as length dilations. We will continue with the example from the last slide. With some minor changes, this again is from Greene's *The Elegant Universe*. The zoom-wheel is now in space and is revolving fast. It is going so fast that Einstein's contraction from special relativity is kicking in. We get curious of the effects and ask Antonia to measure the circumference and Joey to measure the radius of the wheel. It is simple geometry, Circumference = $2\pi R$. Since pi is a constant 3.14159, the ratio of circumference to radius = 6.28 for all normally drawn circles. So Antonia begins to measure the circumference. Pay attention here. The 12-inch ruler in Antonia's hand has shrunk due to dilation. The wheel itself did not contract! If you are saying "wait a minute!" good for you. Let us digress a bit so that we can address this right away. Why has the circumference of the wheel not contracted? For the answer, I was directed to the notes at the end of the book. An explanation goes like this:

". . . to this day there is not universal agreement on a number of subtle aspects of this example . . . you may be puzzled about why the circumference of the ride is not Lorentz contracted in exactly the same way as the ruler . . . we never analyzed the ride when it was at rest . . . Lorentz contraction of the rides circumference would have been relevant only if we compared the properties of the ride when spinning and at rest, but this is a comparison that we did not need."

So clear as mud, but it covers the ground. Let us continue.

Let us pick up from Antonia's measurements. I think we can still make a case. Because her ruler shrunk, she had more 12-inch measurements than when measured at the slower speed. That means that the circumference expanded. Now Joey does not experience a contraction on the radius because the radius is perpendicular to the motion. So what Antonia has found, that the ratio circumference/radius (C/R) is bigger than for the classic circle constant, is in clear violation of classical mathematics.

Einstein's explanation is so simple as to embarrass. The classical circle based on $C = 2\pi R$ works only in two dimensions. If drawn on a warped or curved surface, its spatial relationship will be distorted, and the 2π relationship will generally not work. Two common illustrations of this effect show a two-dimensional circle with radii lines placed on what looks like a horse saddle and the other placed on a sphere. All three have the same radius, but the sphere example has a shorter circumference, and the saddle has a larger circumference. And this is what convinced Einstein to propose that all of space is curved, based on the fact that the C/R ratio of a circle is violated. The flat geometry of the Greeks, golden for thousands of years had been dethroned to the zoom-wheel and curvature, from the perspective of an accelerated observer.

And it would follow that not only does acceleration warp space, but it also warps time.

Actually, this is not as difficult as it might sound. Let's join Antonia and Joey again on the zoom-wheel. Let's give them both a watch. We will place Antonia at the back wall standing up and Joey in the very center. When both are ready, we will ask Joey to slowly crawl toward Antonia. At any given time, Antonia and Joey will both have the same number of revolutions. But Antonia is traveling much faster under the force of acceleration and therefore will experience time dilation—time is slowing much more so than for Joey. When Joey crawls halfway to Antonia, his watch has also slowed but still not nearly as much as Antonia's watch. Warped time, then, means that the passage of time is different at different locations on the wheel.

Einstein previously had made the connection between acceleration and gravity as being indistinguishable. With this last piece of the puzzle, Einstein had made the big leap—*Gravity is the warping of space and time.*

It follows, then, the more massive the object, the greater the warpage of space. It would follow then that the more distant an object gets from a warpage, the weaker the influence. Warpage occurs around all objects including earth and each of us. The agent of gravity then, according to Einstein, is *the fabric of the cosmos.*

Chapter Four

The Big Bang

The Early Moments

A COMMON MISCONCEPTION is that the big bang provides a theory of cosmic origins. It doesn't . . . It tells us nothing about time zero itself. It tells us nothing about what banged, why it banged, how it banged, or frankly whether it ever really banged at all. (*The Fabric of the Cosmo,* Brian Greene)

When Edwin Hubble proved in 1929 that the universe is expanding, Einstein was disheartened knowing that this discovery would have been his if only he believed his own equations twelve years previous. This is considered one of the greatest discoveries of all time. However, this is not the death of the cosmological constant as we will see later. The reason that Einstein wanted to trust his intuition and believe in a static universe most likely is that he and the scientific community would then not have to deal with a beginning and an end to the universe. Simply put, a static universe would have always been and would always be—nice and tidy. Alexander Friedman did believe in an expanding universe and found what is now known as the big bang solution to Einstein's equations.

So it is 1929, and the expansion of the universe has been declared. But with expansion, it now must be asked, "How did the universe begin and how does it end?" In this chapter, we are only going to look at the beginning of the universe. We will then come back to the end, and it won't be glamorous.

First, let's rewind from today to just after zero time. It took 13.7 billion years to get here today from the big bang, so we would expect that amount of time to rewind. We will do it much quicker. Today we are going to imagine that all of the matter in the universe stops expanding and begins to recede. This is hypothetical of course since we now know that the universe will expand indefinitely. The fabric of space begins to shrink, bringing all galaxies closer to each other. Over billions and billions of years, the galaxies compress while at the same time temperature rises dramatically. After billions of more years, the stars and other matter disintegrate into a hot plasma of elementary matter. As the fabric continues to shrink, temperatures skyrocket and density magnifies. We are near zero time. Imagine the amount of matter with 100 billion galaxies each with a 100 billion stars squeezed to astounding density. I want you to understand this, so I paraphrase from Brian Greene, ". . . the entire known universe is compressed to a size that makes the dot at the end of this sentence look gargantuan." (*The Fabric of the Cosmos*, Brian Greene, p. 247-248).

There is no more "space" because the fabric of space has also been shrunk and compressed around the point, known as a singularity. Quantities such as density, temperature, and space-time curvature become mathematically infinite as volume goes to zero. When density goes to infinity, then all of the matter in the universe will fit in this singularity. This is conceptually not easy to absorb, but it is important to the view of the big bang. We need to get into this some more, but let's move on for now.

Now we are going to reenact the big bang. From a point-sized singularity under unimaginable compression and temperature, came an eruption, which spewed out all of space and all of matter—13.7 billion years ago. The temperature of the universe at 10^{-43} seconds after the bang (Planck time) is calculated to be 10 trillion, trillion times hotter than the interior of the sun ($\sim 10^{32}$ degrees Kelvin). The primordial plasma is beginning to form clumps. At about one hundred thousandth of a second, the universe had already "cooled" to a million times hotter than the sun (~ 10 trillion degrees Kelvin). At this point, the quarks (remember the elementary particles?)

are now able to get together in groups of three. At about one hundredth of a second later, the nuclei of some of the smaller elements (protons and neutrons) are beginning to congeal out of the primordial soup. During the next three minutes with the temperature getting down to a billion degrees, nuclei of predominantly hydrogen and helium continue to emerge, along with traces of deuterium and lithium.

Take a deep breath because nothing else really happens for the next few hundred thousand years.

But then, when the temperature dropped to a few thousand degrees, the sea of electrons (another elemental particle) had slowed enough so that positively charged atomic nuclei, mostly hydrogen and helium could capture the negatively charged electrons to produce the first electrically neutral atoms. This was a very significant step in the universe because from this point on, the universe became transparent. This is because photons were very much part of the primordial soup. However, the photons prior to this step interacted with all of the charged particles, which limited them from escaping the hot soup. Now with the free electrons being captured by the free nuclei, the photons for the first time were released from entrapment, and they have been traveling throughout the universe ever since.

Big Bang versus Steady State

With the formation of hydrogen, helium, deuterium, and lithium at around the 300,000-year time frame, scientists disagreed well into the 1960s as to how the other higher elements were formed in a process called nucleosynthesis. At the same time, scientists were debating other models for the origin of the universe. Some objected to the big bang on the basis of material suddenly arising from nothing. I am sure most of us would object on the same basis. The objectors proposed a steady state model. It in turn had its problems. One positive by-product of the steady state theory, put forth in 1957, was a mechanism for the formation of the higher elements. As proposed, these elements developed within and released by catastrophic supernova explosions. It was shown that these stars have high enough temperatures to give birth to all of the elements from small and up. This concept became a landmark in astrophysics and William Fowler received a Nobel Prize for it in 1983. However, the argument for big bang or steady state continued into the 1960s.

The final persuasion for the big bang came in 1965 when Bell Labs in New Jersey scanned the Milky Way for radio emissions with a giant antenna. To their amazement, instead of the expected normal reverberations, they encountered a persistent background hiss, like the echoes of an early universe. "Their unprecedented findings demonstrated that the cosmos is bathed in the cooling afterglow of a searing earlier epoch." (*Brave New Universe*, Paul Halpern and Paul Wesson, p. 91)

That background hiss was first heard by Arno Penzias and Robert Wilson who noted that the noise came from all directions—from everywhere. A Princeton group calculated the temperature of the radiation that would produce such signals. That would be 3° Kelvin (minus 454° F). They then determined the temperature of a fireball cooled by billions of years of expansion to be 3° Kelvin; thus, the final proof that the early universe was once hot and dense. The background hiss was later to be called cosmic microwave background (CMB)—the same waves as your microwave ovens, photons given off by electrons changing to a lower energy orbital in the everyday life of an atom. Small universe!

The winner: Big Bangers!

Penzias and Wilson were awarded the Nobel Prize in physics in 1978.

There are and have been many models for the beginning of the universe, and there will always be dissenters, but for now, the big bang is widely accepted by the scientific community as the leading contender—albeit it's still a theory.

Inflationary Cosmology

For many decades, some of the most basic questions about the big bang model went unanswered. One of them is what happened at zero time? To try to answer this, cosmologists in the 1980s revisited Einstein's infamous cosmological constant. It was intriguing because it was formulated to allow for antigravity or repulsion—a concept more suited for a static universe. But physicists realized that this concept of antigravity in just the right environment could account for the birth mechanism of the big bang—and that environment existed in the beginning of the universe. In Brian Greene's words, "For a time interval that would make a nanosecond seem like an eternity, the early universe provided an arena in which it exerted its repulsive side with a vengeance, driving every region of space away from every other with unrelenting ferocity (authors are unified in not calling it an explosion). So powerful was the repulsive push of gravity that not only was the bang identified, it was revealed to be bigger, much bigger than anyone had previously imagined. In the inflationary framework, the early universe expanded by an astonishingly huge factor compared with what is predicted by the standard big bang theory, enlarging our cosmological vista to a degree that dwarfed last century's realization that ours is but one galaxy among hundreds of billions."

So Einstein's cosmological constant reemerged from mathematical limbo to take on this new role "and ushered in one of the most dramatic upheavals in cosmological thinking since our species first engaged in cosmological thought." (*The Fabric of the Universe*, Greene, pp. 272-273)

Formation of Stars

So the lighter elements had been formed just after 300,000 years from the initiation of the big bang, thus freeing photons from their captivity and allowing them to roam the universe forever. This has become the background radiation of the cosmos, a testament to its beginning. The higher elements were formed from supernova explosions probably later in the time frame of 100 million to 1 billion years ago. Other matter appeared as clumps. Atomic hydrogen formed into molecular hydrogen (H_2) and then into a molecular clouds and ultimately nebulae, which were precursors to star formation. The first star was probably a monster in size. But since these bigger stars have a shorter life span, they would not be around today. Stars are gigantic furnaces fueled by (as you know) hydrogen. In a process called fusion, hydrogen plus hydrogen fuses into helium plus energy—massive amounts of fireball energy. When the star uses all of its supply of hydrogen, it begins to convert helium into oxygen and carbon. If the star is massive enough, it will continue converting oxygen and carbon into neon, sodium, magnesium, sulfur, and silicon. (Lucky for us that not all stars are massive enough for them to continue converting oxygen and carbon). Eventually all of the elements will be converted into calcium, nickel, chromium, copper, and others until iron is formed. This stops the sequence because the star is not hot enough to fuse iron. The inward pressure of gravity wins out over the outward pressure of the nuclear reaction, and the star collapses on itself.

Our sun is but one of about 100 billion stars in the Milky Way alone. Scientists believe that there are about 10^{22} stars in the universe—more stars than grains of sand on all the beaches on earth. For this reason, hydrogen and helium are the two most abundant elements in the universe. They are a part of what is known as the visible universe. But let's not get ahead of ourselves. Let's go back to the formation of hydrogen and the molecular cloud mentioned above and try to put the evolution of celestial matter in some sort of time sequence. However, keep in mind that much of what is happening here is happening at the same time—at least as cosmic time will allow.

Formation of Celestial Matter

Refer to illustration 7. Nebulas are cosmic occurrences of gas and dust in space. As such, they contain the elements from which stars, solar systems, and galaxies are built. That would be about 90% hydrogen, 10% helium, and 0.1% traces of heavier elements. They can be quite expansive of up to several hundred light-years across. The trigger to star formation from a nebula may come from the gravity of a nearby star or a shock wave from nearby supernova. Over a long period, mass coalesces until it achieves a critical mass and begins to heat up. With continued heat, eventually, it achieves a critical mass and fusion begins. The life of a star depends upon its size. Larger stars burn much faster and may have a lifetime of only a few hundred thousand years, smaller ones for billions of years. When a star gets near to using up all of its hydrogen, they will expand into a supergiant in the case of a massive star and a red giant in the case of normal star. They will continue in this state until all available fuel is exhausted, when gravity takes over, and the star collapses. Most average stars will blow away their outer atmosphere and join the ranks of nebulas. Their core will remain and burn as a white dwarf until it cools down and turns into a black ball of matter known as a black dwarf (not to be confused with a black hole). If the super giant is big enough, the collapse will be of such proportion as to trigger a massive explosion known as a supernova. If the remaining mass is greater than 1.4 times that of the sun, it will collapse further into a neutron star. At this stage, the atoms are compacted into a dense shell of neutrons. If the remaining mass of the star is 3 times that of the sun, the star will collapse into a black hole—never to be seen again (more later).

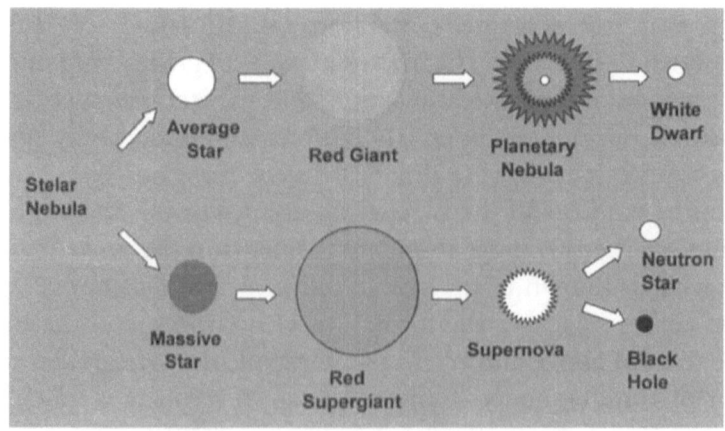

Illustration 7. Life Cycle of a Star

Pulsars

Pulsars were first observed in 1967 while scanning the skies for stars. It was a star-like object that emitted radio waves as pulses. While radio waves are not uncommon in the cosmos, this is the first time that they were observed in very tight pulses, in this case one per second. Bright pulsars have been observed in just about all frequencies including visible light.

It has now been determined that a pulsar is actually a rapidly spinning neutron star, which if you recall is the highly compacted core of a dead star left behind in a supernova explosion. They have powerful magnetic fields that can be up to one trillion times that of earth. If the neutron star is aligned properly, the powerful magnetic rays will flash across earth with every pulse, effectively much like a lighthouse beam of light. Different pulsars pulse at different rates depending upon the size and mass of the neutron star.

Pulsars are found by using large radio telescopes. Recently, the 1,000th pulsar was found by a radio telescope located in Australia. Pulsars are incredibly accurate timekeepers, more accurate than the best atomic clock, currently the most accurate timekeeper on earth.

Quasars

Quasars are star-like objects that even today astronomers are mystified by what these objects are. They are very bright and very distant, well beyond our own galaxy. They emit enormous amounts of energy and can burn with the intensity of a trillion suns. In fact, some quasars are believed to burn out 10 to 100 times the energy of our entire galaxy—wow!

Quasars are so distant that the light that we see when we observe them today have been traveling for billions of years. The most distant of the quasars observed are 10 billion light-years away—we are seeing what they looked like 10 billion years ago! They are, in fact, the most distant objects to ever to be detected in the universe. They may not even be around at this point.

Not to be left out, the speed of quasars is extremely high. By the red shift, quasars are moving at about 240,000 km/sec (about 150,000 mi/sec) or about 80% the speed of light. Think about the relativistic effects!

Today more than 2,000 quasars have been identified.

Black Holes

A black hole is a remnant of a star cycle that begins with a massive star that expands into a supergiant that collapses into a supernova, which collapses into a neutron star that implodes into . . . it. All of the matter that was left over in the neutron star is now a single point in space that is so massive that mathematics go haywire. The laws of physics and the fabric of space seem to break down and cease to exist—volume becomes zero, density becomes infinite, and time stops. So are they real? Yes, although no one has seen one, and no one will ever see one because they are so dense that even light can't escape. Then how can astronomers be sure? Well, they look for two signs: (1) areas of space that exhibit a large amount of mass in a small, dark space, (2) strong amounts of x-rays. Over the years, astronomers are convinced that black holes are real and not so far away.

Astronomers now believe that a black hole exists right in the center of our galaxy. In fact, they believe black holes exist in the center of many galaxies, as well as one of two binary stars. Since many stars in the universe are binary, occasionally, one of the two goes through the star cycle and end up with a black hole. This is particularly helpful because the black hole will begin to suck matter off of the star and send it into orbit surrounding the black hole. It is known as an acceleration disk. As this matter gets sucked into the black hole, it emits very strong amounts of x-rays.

Upon finding such a binary star, astronomers can determine the orbital speed of the visible star and in turn determine the mass of the invisible companion. If the mass is high enough, then it most probably is a black hole.

Planets

Planets and galaxies fall generally under a time-frame window of 1 billion to 12 billion years after the big bang. In the scheme of things, I am not sure in what order earth, our solar system, and our galaxy were put together. Their formation is still somewhat of a mystery. One hypothesis has it that the leftover gas and dust from star formation forms a ring around the star. Ultimately, gravity causes lumps to form, and over a long period, the process of agglomeration builds them into a spherical shape. In more time, the lumps grow by the process of collision and merging. Gravity causes the heavier elements to separate from the lighter ones to form a core and a crust and a gas atmosphere.

The existence of other planets around other stars has been suspected since 1916. The first visual proof of a planet beyond our solar system was in 1999. Today there is considerable effort to identify a planet or planets with conditions similar to earth to determine if other life-forms have evolved.

In chapter 5, we will spend considerable time on planet Earth.

Galaxies

Galaxies are huge accumulations of matter in all the forms just reviewed. They are works in progress as the universe is in constant change. As mentioned before, the universe is commonly described as consisting of 100 billion stars in our single Milky Way galaxy and 100 billion galaxies in our universe. I presume there is a lot of scientific license in these numbers. One source indicates recent estimates in our Milky Way of 400 billion stars. Another source estimates galaxy formation beginning about one billion years ago, yet a press release dated 2004 indicates that Milky Way formation initiated about 200-300 million years after the big bang, about the same time as photon was released into the universe shortly after the big bang. That would put galaxy origins at about 13.4 billion years, just after the big bang. So galaxy formation comes much, much earlier than earth formation, which was 9.4 billion years after the big bang.

Our galaxy is a gigantic-type spiral disk with a bright center bulge. As galaxies go, the Milky Way is a giant with a diameter of 100,000 light-years (one light-year is 6 trillion miles). Our solar system is located ¾ away out from the center (about 28,000 light-years) in one of its spiral arms. Our solar system rotates around the galactic center at a speed of about 150 mi/sec and requires 220 million years to complete one orbit.

The size of galaxies and the distance between them are difficult to fathom. And the dynamics are that all of the galaxies in the universe are moving away from each other at tremendous speeds. The farthest galaxies discovered so far are 10 billion light-years away from earth. In order to get a better picture of the universe, scientists are now working on a project to plot the locations of millions of galaxies. Curiously, galaxies tend to belong in clusters. Our Milky Way, as well as nearby (in cosmological distances) Andromeda, along with a bunch of other galaxies belong to the Virgo supercluster. Most clusters are much bigger. Interestingly, what appeared once to be a random arrangement is now beginning to take on a complicated design. Stay tuned.

Our Solar System

Whereas our galaxy began to form 200-300 million years after the big bang, our solar system emerged from the gravitational collapse of a giant molecular cloud 4.6 billion years ago or 9.1 billion years after the big bang. Earth emerged about 300,000 years after that, 4.3 billion years ago or 9.4 years after the big bang. The various planets formed from the remaining cloud of dust and gas. They grew from grain size to clumps to larger bodies and over a few million years into their current planetary size. The inner planets were too warm for volatiles like water and methane and therefore were smaller and left with mostly high melting point compounds like silicates and metals. Farther out, planets like Jupiter and Saturn, where more volatile, icy compounds could remain solid, became the gas giants. Still colder planets like Uranus and Neptune became ice giants with cores that are believed to be mostly ice. Earth circles the sun at a speed of 66,600 mi/sec.

On a star-brightness scale, our sun is about right in the middle of the other suns in the universe. It is in the prime of its life right now with plenty of hydrogen fuel left. This means that it is growing brighter with time. In case you are wondering, yes, the sun will ultimately turn into a red star, and earth will turn into an ice cube. But not to worry, this won't happen for a long, long time. The sun's cycle is estimated to be half-way through. It will eventually become larger, brighter, cooler and redder. Ultimately, it will become a red giant in 5 billion years. We will see later, the consequences of this death spiral have significant portents for earth much earlier than 5 billion years.

Dark Matter

This chapter has taken us from the big bang to the building of atoms and molecules and to the birth of stars and other celestial matter. Now we are going to muddy the waters a bit. It turns out that scientists have been able to calculate the total mass of the universe, within limits of course, and have calculated the amount of mass required to hold the universe together by the force of gravity. It turns out that the visible mass in the universe is only 5% of that number. It appears that what we see is not all there is. Strong evidence indicates that 95% of matter is invisible and is referred to as dark matter and dark energy. Evidence for dark matter is its gravitational influence on visible bodies. Also, some galaxies rotate much too fast for stability. They should literally be flying apart and begs the question, is dark matter somehow stabilizing these systems? And thirdly, dark matter can evidence itself by bending the path of light.

Certainly some of the dark matter is made up of ordinary baryonic mass (protons and neutrons) that is non-fusion based—planets, moons, brown dwarfs, dust clouds, white dwarfs, neutron stars, and black holes. These are sometimes referred to in the scientific community as MACHOs for massive compact halo objects. However, visible matter and dark matter together only comprise about 30% of the total matter required to hold the universe together. Scientists believe that there is much more matter out there that may consist of non-baryonic "exotic" matter such as neutrinos, axions, super-symmetric dark matter, and WIMPs for weakly interacting massive particles. This whole subject is still a mystery to scientists.

This is an interesting story. In 2004, astronomers in England did a radio survey of the Virgo cluster. They found a disk of hydrogen atoms approximately 100 million times the mass of the sun. When they gauged the rotational speed of the entity, it presented itself as being 1,000 times heavier or 100 billions times the sun's mass. The region was checked by telescope and found nothing visible. They concluded obviously that the extra weight was unseen weight—invisible. 99.9% of the total weight was invisible. This is now recognized as the first-ever, starless galaxy. (*Brave New Universe,* Paul Halpern and Paul Wesson *Chapter 4*)

Horizon Problem

The beginning of inflationary cosmology was the big bang itself. Recall that a tiny singularity containing all the matter in the universe became incredibly hot and dense, and mathematics became meaningless. It triggered an enormous repulsion, an antigravity, initiating the big bang. This mechanism of antigravity was first formulated by Einstein as the cosmological constant that would help to support his notion of a static universe. This same cosmological constant was revived in 1979 (that was not that long ago!) by Alan Guth to help him develop the theory of the big bang. What made this such an extraordinary stroke of genius is that Guth recognized that the cosmological repulsion was a much, much bigger force than Einstein realized; one that would explain the inflationary concept of the big bang. Hang on to this thought.

The inflationary model also has served to provide an answer to another problem, called the *horizontal problem*. This concerns the uniformity of the microwave background radiation (MBR) that is known to be 3° Kelvin +/- 0.001° (minus 454° Fahrenheit) throughout the universe. This has been stumping scientists since the identification of MBR in 1965 because they could make a better case for nonuniformity. But again, the very early big bang provides a satisfactory explanation for the uniformity.

Flatness Problem

A second problem addressed by inflationary cosmology has to do with the shape of space. There are three ways that the universe can curve, making certain assumptions: (1) positive curvature such as a sphere, (2) negative curvature such as the shape of a saddle and, (3) zero curvature such as an infinitely flat surface. From general relativity, it is known that the choice of curvature is dependent upon the amount of mass/energy density per volume. The critical density is that which is just between the two extremes and is 5 hydrogen atoms per cubic meter (about 10^{-23} grams)—this is flat and infinite. Relativity equations show that if the critical density is exactly maintained early on in inflation, then it would be maintained throughout expansion. But if the critical density is either slightly high or slightly low, the subsequent expansion would be today very high or very low. Calculations of the critical density are still in the process after a couple of decades, but it can be said that it is not flat, but it also is not what the above calculations predicted. The matter/energy density one second after the big bang needed to have been within a millionth of a millionth of a percent of the critical density. Any deviation greater than that would mean the matter/energy density value would be way off of today's real value. This is called the flatness problem. This does not say that standard big bang is wrong, but it does make scientists and mathematicians very uncomfortable.

Where Exactly Do We Stand?

We are missing 70% of the dark matter, and inflationary cosmology has not yet revealed to us the fate of the universe. We are going to tie some things together and address these issues. This is interesting—this is important—and this is exciting.

First a quick review! The big bang *is not* an explosion, *it is an inflation*. The wording is important because inflation means that the mechanism is repulsion and anti-gravitational due to Einstein's cosmological constant that Alan Guth resurrected from mathematical limbo in 1980. He realized the tremendous force potential and modeled it into the big bang. This is the inflationary cosmology.

There is one more important step here. In the 1990s, two teams of astronomers, one from Lawrence Berkeley and one from Australia set out to gather the total mass/energy of the universe by measuring the recession speed of type 1a supernovas. Supernovas are massive explosions caused by a white dwarf with a very specific mass. So its explosion brightness is always the same from episode to episode. They serve as lighthouses in a fog—a reference point in space. After locating four dozen of these supernovas and calculating their distances and recessional velocities, both groups came to a totally unexpected conclusion:

"Ever since the universe was about 7 billion years old, its expansion rate has not been decelerating. Instead, the expansion rate has been speeding up." (*The Fabric of the Cosmos*, Brian Greene, p. 299) They concluded that the expansion of the universe slowed down for the first 7 billion years after the initial outburst, before this accelerated expansion.

So the 7 billion year deceleration was expected, the acceleration was not. Why was deceleration expected? Because, ordinary matter and energy give rise to ordinary gravity. But as the universe gets bigger, gravity weakens even though more mass is created. But as gravitational pull is diminished, the repulsive push of the cosmological constant (whose strength does not change as matter spreads out) begins a new era of acceleration.

The discovery by these two astronomy groups has held up to scrutiny through the late 1990s. Now get this, recall the groups of scientists that were measuring the recession speeds of supernovas. Recession speed depends on the difference between the inward gravitational pull of ordinary matter and the outward gravitational push of the "dark energy" supplied by the cosmological constant. Taking the amount of matter, both visible and dark, to be about 30% of the critical density, the supernova researchers concluded that the accelerated expansion they had observed required an outward push of a cosmological constant whose dark energy contributes about 70% of the critical density. Bingo! Eureka! Remarkable!

If this is so, then ordinary visible matter (protons, neutrons, electrons, etc.) constitutes *only 5%* of the mass/energy of the universe! Another 25% goes to some unidentifiable form of dark matter and the other 70% to a totally different and very mysterious form of *dark energy*.

Doesn't this put a totally new perspective on things? There will be one more major surprise in chapter 5.

Chapter Five

The Birth of Earth
The Evolution of Life
The End of the World as we Know it

Earth is Born

A LL RIGHT, IT is now about 9.4 billion years after the big bang. Our Milky Way galaxy has been in the process of forming for almost this entire time. Our sun emerged before the planets about 6-7 billion years ago. Earth and the planets were formed about 4.5 billion years ago. Our sun became one of roughly 100 billion stars in the Milky Way galaxy. It evolved according to the standard star model discussed in chapter 4. Once the sun completed its evolutionary cycle, it began to exert its gravitational muscle on dust-sized matter consisting of heavier elements. In a long, long process, matter condensed, accumulated, and collided to form clumps, chunks, and boulders. Collisions between larger bodies produced energy in the form of heat, producing molten spheres. In all, nine planets were formed (depending upon who you are talking to) with earth being the third closest to the sun. In the process of planet formation, a lot of junk matter, including asteroids and meteorites, emerged. At one point in the process, a rather large chunk of matter struck earth on an off-center blow and sprayed a large area of earths crust into space. Through gravity and overtime, earth was able to reclaim its spherical shape. The deflected chunk of matter from the collision is still circling earth today—it is the moon. The cratered surface that we can see without a telescope on a clear evening is a testament to its violent birth.

The Hadean (4.5 to 3.8 Billion Years Ago)
Earth was born as a molten fireball. There were no oceans and no oxygen in the atmosphere. During the molten stage, there was a separation of

elements by size and by melting temperature. The heavier elements, such as uranium, gold, silver, and molten iron sank toward earth's center, and lighter elements rose to the surface. It is this stage that created earth's magnetic field with iron at its core. There was no atmosphere at this point.

The Archean (3.8 to 2.5 Billion Years Ago)

About 700 million years later, Earth's surface cooled and had by this time become a planet with an atmosphere and plenty of groundwater—lots of water making oceans and seas. It is thought that the atmosphere consisted of methane, water vapor, ammonia, carbon dioxide, and nitrogen. It was still too early for oxygen in the atmosphere since most of the oxygen bound quickly with hydrogen to form water and with elements to form ores.

The Archean was a time in which the earliest life-forms first appeared on earth. These life-forms were nothing more than one-celled creatures and simple bacterial forms that were able to survive the toxic atmosphere at that time. In fact, our oldest fossils date back to 3.5 billion years. This period right after earth solidified is also known as the beginning of geological history. Earth rocks have been dated as far back as 3.8 billion years. How do scientists know the age of these creatures? Well, these fossils are embedded in rock. Rocks can be dated by radioactive elemental techniques.

Also in this time, about 70% of the world's landmasses were created.

The Proterozoic (2.5 to 0.5 Billion Years Ago)

This period is noteworthy because it was the start of photosynthesis—the ability of plants to use energy from the sun to convert carbon dioxide into carbohydrates (sugar, starch, and cellulose), with oxygen being a by-product of this process. The first fossils of this period were blue-green algae that could photosynthesize. Free oxygen began to build up around 1.8 billion years ago providing the stimulus for oxygen-breathing animals and human life-forms as we know them today and allowing the life-forms striving in the toxic air of that time to die out. It was a huge evolutionary step.

The continents began from landmasses. They stabilized into one supercontinent about 1.1 billion years ago.

The age of humans is about to begin!

The Holocene (500 Million Years Ago to Today)

This was a period of wild activity including the emergence of modern man. There were four extinguishing events between 60 million years ago and 570 million years ago that are thought to have been two collision events, one ice event and one volcano event. Some resulted in major kills of most of the life-forms at the time. Could mankind today survive such an event or events? After the dinosaur extinction 60 million years ago, things quieted down, at least from a bombardment viewpoint. However, earth will still see ice ages that we will talk about later. But the table seems to have been set for the emergence of humankind.

The Evolution of Life

You can now download from the internet the near-complete instructions for how to build and run a human body.
Genome—The Autobiography of a Species in 23 Chapters, Matt Ridley

Recall that life-forms existed as far back as 3.8 billion years, and at that time, oxygen was not part of the atmosphere. So the energy for these early life-forms life came from organic chemicals. So question number one is "how did the first single cell life-form evolve?" How did simple non-living chemicals react with each other in way that allowed it to replicate—to copy itself, for that is the basic prerequisite for life. This feature alone allows a grouping of chemicals to propagate and grow from simple to complex.

The outer membrane of that first cell was probably as we know it today and is simple considering the natural chemicals available at the time. However, the nature of this replicator molecule at the time, according to Matt Ridley, is unknown. But it is assumed that by trial and error over time, those replicators that worked better at copying "survived." Possible chemicals that could have served as agents for a replicator and cell membrane are proteins, nucleic acids, and phospholipids. These are the basic chemicals of life. They could have been formed from available organics like ammonia and methane commonly found in the atmosphere at the time. What would initiate the reaction? Well certainly there are numerous sources of energy that would include volcanoes (on land or below the sea), lightning and ultraviolet radiation. Throw in time (lots of time), chance, and probability, and chemistries can, and did, go rampant.

The early cells, then, transformed from using organic molecules as an energy source to using sunlight as an energy source. They used carbon dioxide to form polysaccharides. This happened about 3 billion years ago with a process called photosynthesis. Since oxygen was a by-product of this process, it eventually accumulated in the atmosphere. In doing so, it set the stage for animal and human evolution. A secondary benefit of the emergence of oxygen in the atmosphere is that its ozone form (O_3) acted like a filter for the otherwise damaging effects of ultraviolet radiation from the sun. The cellular organisms that migrated to Earth's surface were less likely to die. Around 2.6 billion years ago, such organisms became adapted on land. However, water was more protective of ultraviolet radiation, and fish ultimately filled the oceans—until 530 million years ago. In that period, the first of a number of cataclysmic events occurred. It wiped out most of the life-forms thriving at that time.

But some life-forms survived. At about 488 million years ago, land plants again thrived, and the oceans were again filled with fish 450 million years ago. And there is evidence that the water life-forms made the conversion to land life-forms around the same time, thriving on the land plants. At 440 million years ago, another extinction event occurred and was thought to be an ice age. That was followed by another event about 365 million years ago. Twenty-five million years later, there is evidence for land egg-laying species. Ten million years later, reptiles, birds, and mammals had emerged from amphibians. Now here we go again. The most severe extinction event took place about 250 million years ago with an estimated 95% kill rate, thought due to a volcanic eruption. But life was again resilient. Twenty million years later, dinosaurs and all their friends appeared. There was an extinction event around 200 million years ago, but the dinosaurs were spared. However the dinosaur's existence had run out of luck. Sixty five million years ago, Earth was struck by a 10-kilometer meteorite just off the Yucatan Peninsula. The sky was filled with debris, and photosynthesis ceased. Most of the large animals became extinct. But we don't need to feel too bad for the dinosaurs since their species existed for about 170 million years. Humankind is working on a paltry 200,000 years of existence.

The Main Event, Human Species

"A small African ape living around 6 million years ago was the last animal whose descendants would include both modern humans, and their closest relatives, the bonobos, and chimpanzees." (*http://en.wikipedia.org/wiki/History_of_Earth*, page 7)

Its family tree has only two branches, one of which developed the ability to walk upright. About 2 million years ago, brain size had increased substantially, and the species became the first to be classified in the genus *Homo.* The ability of *Homo* to control fire was developed in the time frame of 800,000 to 1.5 million years ago. Over the years, brain size continued to increase, learning capacity increased, social skills improved, language skills advanced, and tools became more elaborate. Modern humans, *Homo sapiens,* arrived in Africa about 200,000 years ago. So you see that no humans were around during the dinosaur era 230 million to 65 million years ago despite what you see in some of the older classic movies. The first species to exhibit spirituality were the Neanderthals who buried their dead perhaps in hope of an afterlife. Then about 32,000 years ago, the Cro-Magnon cave paintings were thought to have some religious significance. By 11,000 years ago, *Homo sapiens* had reached the tip of South America thus inhabiting all of the continents.

Homo Sapiens **(Appeared 200,000 Years Ago)**

Homo sapiens were hunter gatherers, whose fictionalized stories were so interestingly captured by Jean Auel's *The Mammoth Hunters, Clan of the Cave Bear* and others. It was way before recorded history. Somewhere around 8500 BC to 7000 BC began farming and the taming of animals. This led to the first civilization around 4000 BC to 3000 BC in the Middle East, followed by other civilizations in Egypt and the Indus River Valley. Starting about 3000 BC, the oldest religion, Hinduism, was initiated and is still practiced today. Writing was invented and henceforth the world's history has been documented as libraries stored the information. This allowed for the pursuit of knowledge for the first time. Science arose as a discipline. Civilizations flourished and trading began.

Around 500 BC, civilizations turned into empires—the Middle East, China, India, Iran, and Greece. And then the scourge of humankind was invented as the empires took to war in the never-ending struggle for space, resources, riches, power, and even simple differences. Wars and conquests have been a part of human nature ever since. They include minor skirmishes as well as two world wars, one from 1914-1918 and the other from 1939-

1945. The League of Nations has given way to the United Nations as a forum for world leaders. The Greek civilization has given way to the Roman Empire, to the Renaissance, to the Inquisition, to the Reformation, to the Age of Reason, to the Age of Science, to the Industrial Age, and to the Age of High-Technology. The latter includes nuclear weapons, nuclear energy, computers, wireless communication, nanotechnology, globalization, genetic engineering, stem-cell research, and others. Space has been conquered, and man has landed on the moon. Since the year 2000, there has been a continuous presence of humans in space. Where do we go from here? Who knows? Human issues include poverty, infectious diseases, war, famine, terrorism, global warming, alternate energy, and too many others.

The major evolutionary milestone events on Earth are provided as follows:

- 4.3 billion years ago: Earth is a molten fireball.
- 3.8 billion years ago: Earth had cooled, the earliest simple life-forms had developed, water had filled the seas and oceans, and an atmosphere had developed but is devoid of oxygen.
- 1.8 billion years ago: Algae forms and photosynthesis begins. Oxygen is released into the atmosphere as part of their life cycle. This is a huge step because it paved the way for the evolution of animals and human life-forms as we know them today.
- 600 million years ago: Life species exploded into many forms in a very short time developing all the major groups of plants and animals of sea and land.
- 500-60 million years ago: At least four massive extinction events killed much of the species of the time. But new species sprung up again every time and very rapidly, as if testing for heartier species.
- 6 million year ago: The last animal appears whose descendants would include modern humans.
- 2-6 million years ago: Humans developed the ability to walk upright.
- 2 million years ago: Their brain size increased substantially.
- 0.8-1.5 millions of years ago: Man controlled fire.
- 200000 years ago: *Homo sapiens* arrived.
- 11 thousand years ago: *Homo sapiens* spread throughout the world.
- 10 thousand years to today: Modern man emerged.

To put in proper time perspective, consider that the emergence of modern humans just 10,000 years ago on an earth that is 4.3 billion years old. Imagine what would happen if we compress 4.3 billion years onto a one-year calendar. This amounts to humans arriving for the New Year's Eve party at 12:59:57 p.m. In essence, humans arrived with three seconds left, barely time for a New Year's Eve kiss!

Human Chemistry

"All life is chemistry." Jan Baptista van Helmont, 1684

The human body is 99% made up of six elements. Oxygen is by far the largest at 65%. This is because the human body is 50-70% water (depending upon age) and oxygen is 89% the weight of water. The other five in order are carbon, 18%; hydrogen, 10%; nitrogen, 3%; calcium, 1.5%; and phosphorous, 0.35%. The top four elements make up 96% of body weight. Note that phosphorous exists at only 0.35% of body weight, but it plays a critical role in the chemistry of DNA and RNA. Actually, there are 28 total elements in the body, and I list them here for your information. In addition to the top six, the following 22 elements exist: potassium, sulfur, sodium, magnesium, copper, zinc, selenium, molybdenum, fluorine, chlorine, iodine, manganese, cobalt, iron, lithium, strontium, aluminum, silicon, lead, vanadium, arsenic, and bromine. Although all of them are important to the function of the body, individually, all are less than 1% of body weight. The last eight are so low they are known as trace elements and are measured in parts per million. Did you make note of lead and arsenic? It is interesting that all life-forms, from the lowest bacteria to the largest of African animals are composed of the same six elements.

Human Genetics

. . . Life consists of the interplay of two kinds of chemicals—protein and DNA. Protein represents chemistry, living, breathing metabolism and behavior—what biologists call phenotype—DNA represents information, replication, breeding, sex—what biologists call genotype—neither can exist without the other. (*Genome,* Matt Ridley)

The normal human body consists of about one hundred trillion cells. Each cell is approximately one-tenth of a millimeter long and is visible to the naked eye. Refer to the visual, illustration 8. Inside each cell are 23 pairs of chromosomes, 46 in all. In each pair, one was contributed by the father and the other by the mother. Therefore each chromosome in a pair is slightly different. Each of the 23 human chromosome pairs carries DNA molecules, and the "instructions" differ in all 23. This is the key to the uniqueness of each human individual. They are designed to build a human being, each chromosome to a different part of the human anatomy.

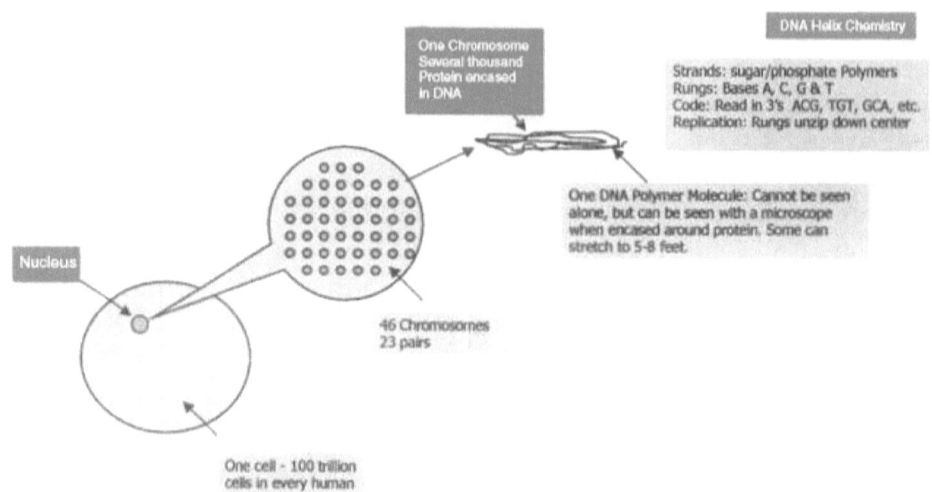

Each chromosome also differs in size. The scientific community has assigne the first 22 pairs in order of size, 1 being the largest. The remaining pair, number 23, carries the sex chromosomes. Women have two X chromosomes, an men have X and a Y chromosomes. The Y chromosome is the smallest of all (damn!!) , and the X falls in between 7 and 8. A single chromosome is a conglomeration of DNA wrapped around thousands of different protein molecules. Chromosomes can be seen through a microscope.

Illustration 8. From Cells to DNA

Each chromosome also differs in size. The scientific community has assigned the first 22 pairs in order of size, 1 being the largest. The remaining pair, number 23, carries the sex chromosomes. Women have two X chromosomes, and men have an X and a Y chromosome. The Y chromosome is the smallest of all (damn!), and the X falls in between 7 and 8. A single chromosome is a conglomeration of DNA wrapped around thousands of different protein molecules. Chromosomes can be seen through a microscope.

So what is DNA? To get started we need to get into some basic chemistry. DNA stands for deoxyribonucleic acid, and RNA stands for ribonucleic acid. In their unreacted states, they are simply sugars not unlike glucose, sucrose, fructose, lactose, etc. Just add ribose and deoxyribose to the list. Their molecular structures can be seen on illustration 9.

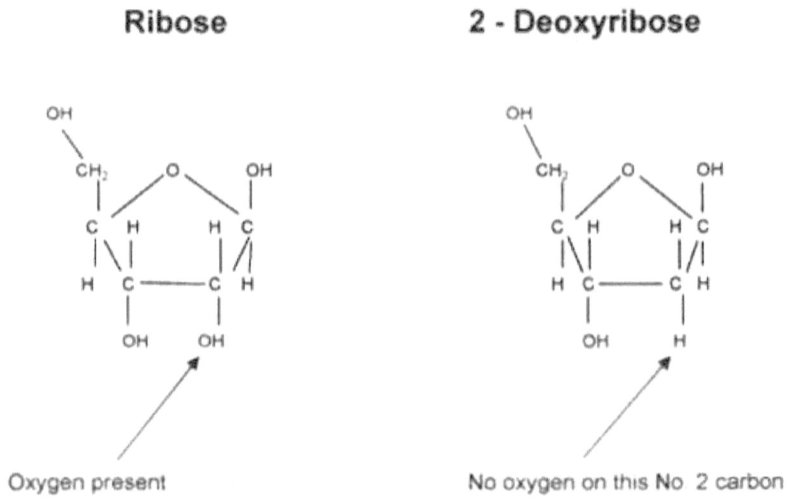

Ribose

2 - Deoxyribose

Oxygen present

No oxygen on this No 2 carbon

Illustration 9. Molecular Form of Ribose and 2-Deoxyribose

When either of these two sugars is reacted with a phosphate group in repeating sequences to form a long chain polymer, we have the beginning step to making DNA. The chemistry of this phosphate/sugar base pair is circled on illustration 10.

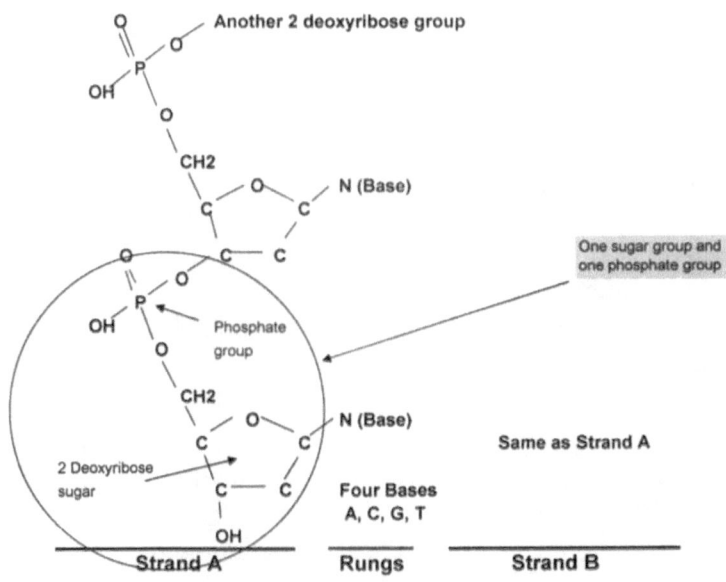

Up to 2+ Million repeating units

Illustration 10. Chemical Structure of DNA

One phosphate/sugar reactant is called a base pair. We will come back to this illustration later. The length of this polymer varies widely and may exceed 2 million segments. See illustration 11.

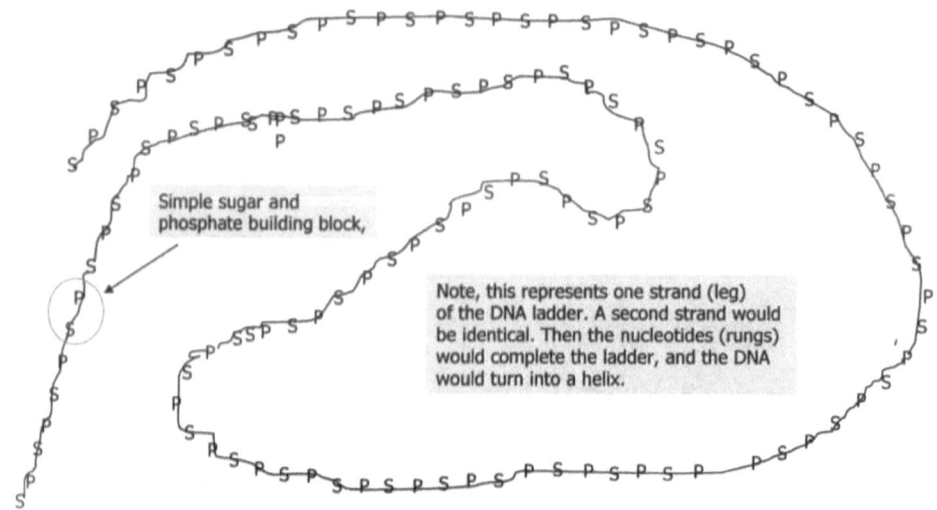

Simple sugar and phosphate building block,

Note, this represents one strand (leg) of the DNA ladder. A second strand would be identical. Then the nucleotides (rungs) would complete the ladder, and the DNA would turn into a helix.

Illustration 11. Simple DNA—One Strand Only

Now the sugar/phosphate polymer that we put together is actually going to be one "leg" of a ladder. That's right; DNA is shaped like a ladder. Since ladders have two legs, we need a second identically shaped phosphate/sugar polymer of same length. Next, we need "rungs" to attach to the two legs to lock them in place permanently. We do not want them to unzip. And lastly, we are going to twist the ladder to give it a double-helix shape, shown below.

We can do that because the legs are flexible, and the rungs hold them in place. This double-helix configuration and its ability to replicate were discovered by Francis Crick and James Watson on February 28, 1953, with the proclamation that "we have discovered the secret of life." You will note that thus far, we have been working with four elements: oxygen, carbon, hydrogen, and phosphorous.

All right, it is time to dig in. We know the chemistry of the strands. We need to know the chemistry of the rungs. The rungs are represented by four and only four, different chemicals. They are the following : (1) cytosine "C",

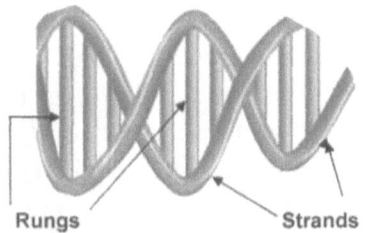

Rungs

Strands

There are four bases - A, C, G, & T. Each rung represents a pair of them coupled in the middle, and bonded to the strands. They couple according to strict laws. The bond in the middle is weakis weak allowing the ladder to "unzip" to give two identical strands ready to rebuild into two identical DNA molecules - thus replicating!

The ribos/phosphate polymer bonds are of great strength. Their length varies from small to greater then 2MM units.

Illustration 12. DNA—The Molecule

(2) guanine "G", (3) thymine "T", and (4) adenine "A" and are referred to as the bases. Out of all the chemicals in the world, these four are crucial for life. T and C belong to a chemical family called pyrimidines that consist of a one-ringed structure. G and A belong to a chemical family called purines that consist of a two-ringed structure. All contain nitrogen as part of their structure. This is important because the nitrogen groups react with the number one carbon group of the sugar on the strands. They form a strong and permanent ester bond.

There are strict rules that come into play with the location and the sequence of the bases because it is the sequences of these four bases that encode information. Some simple laws are the following:

1. There are only four letters (each letter being a base chemical) in the entire code, but because of the length of the strands and endless combination possibilities, they will have the information capacity of 800 good-sized books.
2. The letters work in pairs, A always with T, and C always with G.
3. The genetic code is "read" in groups of three.
4. Sequential groups of three can be thought of as sentences. Let me provide you with some examples.

I would suggest that you go to a DNA visual from the Internet, specifically *http://en.wikipedia.org/wiki/DNA.*

Let us start to build a DNA molecule. Go to the left strand and react the first pair (phosphate/sugar) with base A (adenine). This is totally random; it could be any of the four bases. The reaction is between the number one carbon atom of the sugar to the nitrogen atom of the base. Refer again to illustration 10. Now let's go to the right strand and attach a base. This is not random because the rules say that A must couple with T. So for rung 1, we have A and T. Now for rung 2, there are no rules—start from left or right with any letter. If A is again selected on the left, then again, the rules say T will be on the right. This is all right. In long-chain sequences, repetitions of the same two bases are normal. If G is on left, then C will be on the right. Now the reason for the strict pairing is that A and T combine to give a three-ringed structure (remember pyrimidine = 1 ringed + purine = 2 ringed), and G and C combine for a three-ringed structure. It prevents the possibility of having combined two-ringed or four-ringed structures. The system ensures symmetry.

Now say that we have put together a left rung and a right rung according to the laws. Base pairs are sticking inward toward each other from the strands. Let's say that our polymer ladder now goes out to 1,000 units (pairs). What combines the bases from the left and right rungs? Well, the beauty of this system is that these base molecules join through a mechanism called hydrogen bonding. It is kind of like a handshake. It is an attraction, a very weak bond, zipping A to T, T to A, G to C, and C to G the entire length of 1,000 pair units (they, in fact, may vary from 1,000 to greater than 2 million). It is also the job of the bases to lock the DNA in the double-helix configuration. DNA is not visible, not even with a microscope.

Let us continue with step 3 from above—the genetic code is read in groups of three. Why three? Somewhere in the evolutionary process, this won out as the most efficient process. It is read, for example, GTT, ATG, TAC outward to as many as 700,000 triplets of information. Sequences of information are called "sentences" and ultimately make up what are called the genes.

The central theme of genetics is that the function of DNA is to store information and pass it on to RNA. The function of RNA is to read, decode, and use the information received from DNA to make proteins. The process of replication is division. The replicated DNA then goes with the new cell so that information is preserved. If the cell requires a specific function, such as producing a protein for a specific function, then cell division is not necessary. DNA replication begins in the nucleus with a partial unwinding of the helix by a specific enzyme (protein). This must be done before cell division can occur. The enzymes unzip the DNA at the point of hydrogen bonding, leaving two exact copies of DNA or two identical templates. Free bases move in and position themselves on the strands by the strict rules of A and T, and C and G.

Remember, that your genetic program initiated at conception with one set of genes from your father and another from your mother. Ever since then your DNA, which is filed away in 23 sets of chromosomes, has been busy replicating every part of your anatomy. First, there is a growth function lasting for about twenty years. Hair and teeth mysteriously appear. Muscles develop and hormones are called to duty. Body weight changes on a daily basis. And then we die.

Now before we end, let us go back to the beginning. Try to imagine the very first cell that experienced "life." That would be our most ancient ancestor. After all, all algae are distant relatives. As a part of life, these early forms had to know how to replicate. In doing so, they must have had the same DNA mechanism as we see today. If it did not replicate, by definition, it is not alive. How do we know that these early life-forms constitute life? Its proof is in rocks, 3.5-billion-year-old rocks, with life fossils in them. How is that proof? There are more than one fossil. In fact, there are many, many fossils. They replicated!

"It now seems probable that the very first gene, the 'ur-gene', was a combined replicator-catalyst, a word that consumed the chemicals around it to duplicate itself, and suggests that it may have been RNA. These early 'ribo-organisms' were fragile to heat and size and suffered from gene decay. Over trial and error, and time, a new, tougher form of RNA evolved. A genetic code based on three letters at a time evolved as a more efficient mechanism. One that performed faster and more accurately; it was DNA." (*Genome*, Matt Ridley, p. 18)

And Ridley again (p. 22): "The genes in the cell of your little finger are the descendants of the first replicator molecules; through an unbroken chain of tens of billions of copyings. They come to us today still bearing a digital message that has traces of those earliest struggles of life. If the human genome can tell us things about what happened in the primeval soup, how much more can it tell us about what else happened during the succeeding four million millennia. It is a record of history written in the code for a working machine."

On June 26, 2000, an international consortium of scientists announced that they had completed a rough draft of the complete human genome!

Human Intelligence

Stephen Hawking gave a lecture to a George Washington University audience on April 22, 2008. In response to a question, he conceded that we are probably not alone in the universe but was asked back why we haven't stumbled on to some alien broadcast in space. On the subject of intelligence, he responded, "Primitive life is very common, and intelligent life is fairly rare—some would say it has yet to occur on earth." (*Hawking: Unintelligent Life is Likely on other Planets*, Houston Chronicle, April 22, 2008, Seth Borenstein)

To complete this human story, we must include intelligence. In searching out this subject, I was both amused and surprised at the difficulties inherent in the word *intelligence*. "The evolution of human intelligence refers to a set of theories that attempt to explain how human intelligence evolved. The incomplete nature of scientific knowledge of the human brain is a significant impediment to progress in this field. Another difficulty is that there is no universally accepted definition of intelligence; one definition is 'the ability to reason, plan, solve problems, think abstractly, comprehend ideas and language, and learn.'" (*Hominid Intelligence*, Wikipedia)

Matt Ridley takes a tougher stand. His assessment of the history of intelligence testing is as follows: "There is no accepted definition of intelligence. Is it thinking speed, reasoning ability, memory, vocabulary, mental arithmetic, mental energy or simply the appetite of somebody for intellectual pursuits that mark them out as intelligent?" He concludes that, for the most part, historical efforts have been absurd, crude, bad, and "the stuff of folly". (*Genome*, Matt Ridley, p. 80)

Here again is an interesting story from Ridley. In 1997, Robert Plomin proclaimed that he had found a gene "for intelligence." He had chosen a group of highly gifted twelve- to fourteen-year-old students from the around United States. They had taken exams five years ahead and finished in the upper one percentile. They all had IQs of about 160. From blood samples, he tested for the genes in chromosome 6 and found a different sequence. It turned out to be just slightly different, not always, but often enough to perk interest. To make a long story short, Ridley claims that "it is emphatically not a 'genius gene.'" (*Genome*, Matt Ridley, p. 77)

IQ testing still goes on despite the inherent flaws, and IQ inheritability is still in question. In the 1980s, New Zealander James Lynn noticed that the IQ in all countries was increasing over time, at an average rate of three IQ points per decade. The improvements seem to have nothing to do with what is being taught in school, as opposed to having to do with abstract reasoning ability. This might suggest, according to Lynn, that the effects of the modern world, including sophisticated visual images—cartoons, television, movies, advertisements, and other optical displays—gained at the expense of the written word. This might suggest that these different types of intelligence are at least partly heritable.

Ridley's final point here is that "the environment that a child experiences is as much a consequence of the child's genes as it is of external factors: the child seeks out and creates his or her own environment. If she is of a mechanical bent, she practices mechanical skills; if a bookworm, she seeks out books. The genes may create an appetite, not an aptitude. After all, the high heritability of short-sightedness is accounted for not just by the heritability of eye shape, but the heritability of literate habits. The heritability of intelligence may therefore be about the genetics of nurture, just as much as genetics of nature. (*Genome*, Matt Ridley, p. 90)

Below is a proposed scenario for the evolution of human intelligence, from an article on the internet titled *Hominid Intelligence*.

1. About 10 million years ago earth's climate turned colder and drier, and entered an ice age. Tropical forests in northern Africa retreated and gave way to grasslands and ultimately to desert. Tree dwelling primates gradually adjusted to a ground-dwelling life. This adjustment however made them more vulnerable to their predators. They adjusted again by learning to walk on their hind legs, which gave them longer vision for approaching danger. The forelegs hence evolved hands that would be used for gathering food and making tools and weapons in future years. This development required about 5 million years. The hominid brain was still no larger than brains of mammals of the same size.

2. About 5 million years ago, the hominid brain began to develop rapidly. It is proposed that an *evolutionary loop* had been established between the hands and the brain. Tools gave this species an evolutionary advantage to drive off other hominids not advanced with tools and weapons. They could survive and propagate the species. It was the tools and weapons that required a larger and more sophisticated brain.

3. About 2 million years ago, the hominid brain size continued to grow. The larger brain meant a larger skull, requiring a wider birth canal. This presented another problem: If the birth canal is too wide, then

females would lose their running ability to avoid predators. The evolutionary answer to this was to give birth at an earlier stage of fetal development.

4. This adaptation allowed for the brain to expand further. But this imposed a new restriction on the hominids. Because the fetuses were born so early, they required more care and made the females less mobile. This required that the females cared for the young, and the men hunted and gathered food. Bigger and better tools and weapons were required, with less dependency on size and strength. Each succeeding generation had a larger brain relative to its body and became successively more fine-boned. As we get closer to 200,000 years ago less is said about further increases in brain size. (*Hominid Intelligence, http://en.wikipedia.org/wiki/Hominid_Intelligence*).

Also mentioned in this same Web site is a short summary of an ecological dominance-social competition model explained by Mark V. Flinn, David C. Geary, and Carol V. Ward based mainly on work by Richard D. Alexander. The model proposes that human intelligence was able to evolve due to domination of their habitat. Only then were they "free" to develop more advanced social skills such as communication through language. Their main competition after habitat's dominance shifted from nature to their own species, making it of vital importance to outmaneuver other members of the group through advanced social skills. Refer to: (*http://en.wikipedia.org/wiki/Hominid_Intelligence#Ecological_dominence_Social_Competition_Model*).

Earth and Humans: Comments of General Interest

The diameter of earth is about 8,000 miles, and its circumference is 24,902 miles. Seventy percent of the earth's surface is water; 30% is land. Of the 70% water, 97% is salt water. The highest point is Mt. Everest at 29,028 ft above sea level; the lowest is the Dead Sea at minus 1,392 ft below sea level. The average depth of the oceans is 2 mi. The earth is made up of the following chemical elements: oxygen, 47% (mostly water, H_2O); silicon, 28%; aluminum, 8%; iron, 5%; calcium, 3%; potassium, 3%, sodium, 3%; magnesium, 2%; and others, 1%. Refer to graph 1. In the 1% of others category would be gold, silver, platinum, uranium and all of the other stable elements listed in the table of the elements. Keep in mind that 1% of the weight of Earth is not a small number by any means since the weight of Earth is 6,585,600,000,000,000,000,000 tons.

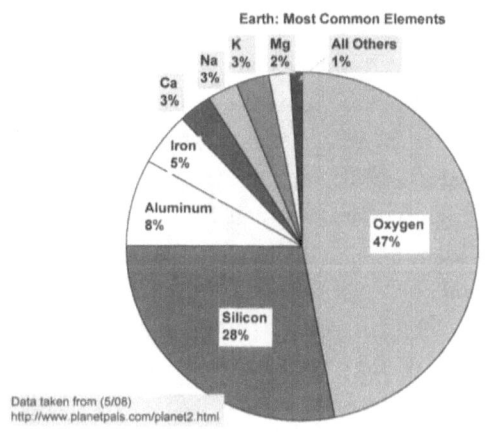

Graph 1. Earth's Elements

Population and Trends

According to GeoHive, the world's population as of May 23, 2008, is 6,669,294,536 (Now that's an exact number!). The top five countries are China (1,329,150,706), India (1,146,060,284), United States (303,537,994), Indonesia (237,214,163), and Brazil (191,707,646). Graph 2 shows the growth in global population from 1950 through 2000 and the forecast through 2050. That forecast number for 2050 is 9,104,000,000. However, the average growth rate through one hundred years has decreased from a peak of just over 2% from 1962 to 1971, to just over 1% in the early 2000s. Projections are for the growth rate to steadily decline to less than 0.5% by 2045. While this is a very positive decline, it still represents an additional 33 million people every ten years.

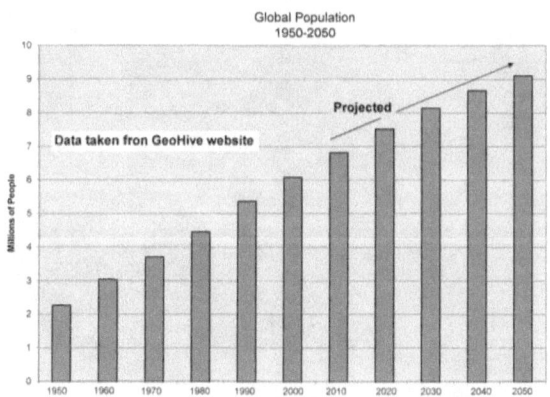

Graph 2. Global Population

Wealth

The ten richest countries in the world as of 2005 are provided on graph 3. Clearly evident is that the United States has truly emerged as a sole superpower with a GDP of $12.5 trillion, followed by Japan at $4.5 trillion, Germany at $2.8 trillion, China at $2.2 trillion, the United Kingdom at $2.2 trillion, France at $2.1 trillion, Italy at $1.7 trillion, Spain at $1.1 trillion, Canada at $1.1 trillion, and Brazil at $0.80 trillion.

However as of mid-2008, the cost of oil has skyrocketed from $75 a barrel last year to $135 a barrel today, with projections for continued significant increases. The obvious result will be some redistribution of wealth to those countries with the most plentiful reserves. More worrisome perhaps is the potential for global instability as the world's powers compete for the available supplies. Effects are not only on gasoline, but on end products that incorporate billions of pounds of plastics and other hydrocarbon based chemicals. Industries that would be affected include automotive parts, paper production, paints, adhesives, rug backing, plastic PVC piping, and home siding and many others that play a quiet but important role in our daily lives. Graphs 4 and 5 provide you with an updated assessment of oil producers and their 2007 reserves. This gets somewhat ominous when you see twelve more years of reserves for the United States and ten more years for Mexico. While we are warned of a good deal of uncertainty in the accuracy of these numbers, I would guess that they are near ballpark. This also precludes future oil findings.

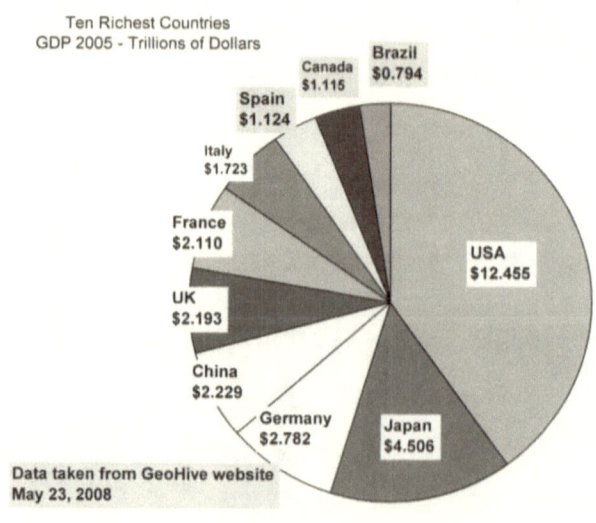

Graph 3. Richest Countries

JERRY MILLER

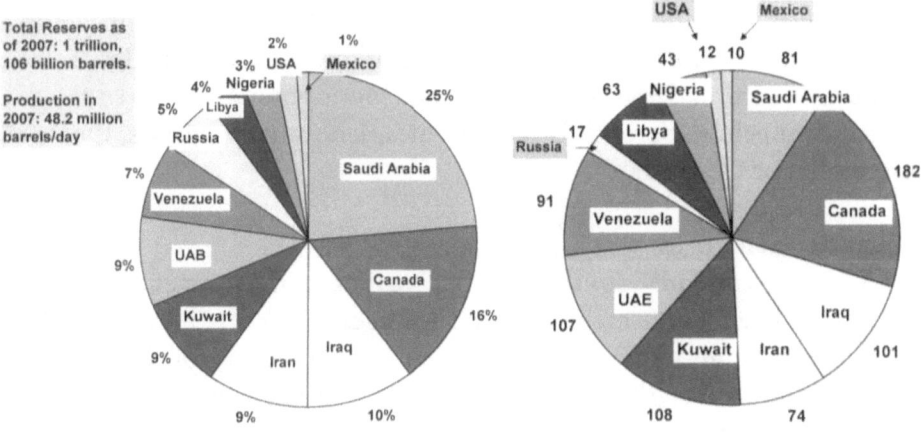

Graphs 4 and 5. Oil Reserves by Country

Christianity	2.1 billion
Islam	1.5 billion
Nonreligeous	1.1 billion
Hinduism	900 million
Chinese Traditional	394 million
Buddhism	376 million
Primal Indiginous	300 million
African Traditional	100 million
Sikhism	23 million
Juche	19 million
Spiritism	15 million
Judaism	14 million
Baha'i	7 million
Jainism	4.2 million
Shinto	4 million
Cao Dai	4 million
Zoroasttrianism	2.6 million
Tenrikyo	2 million
Neo-Paganism	1 million
Unitarian	800 thousand
Rastafarianism	600 thousand
Scientology	500 thousand

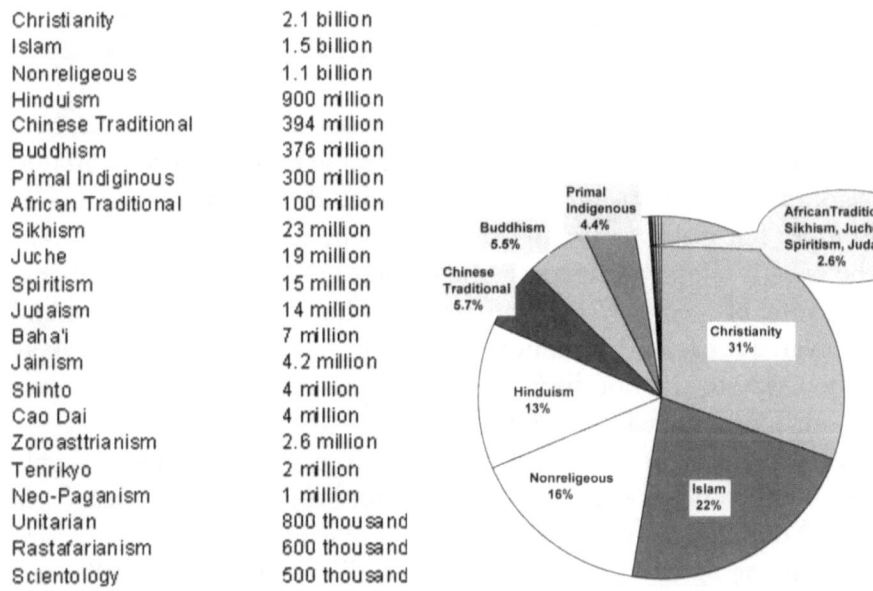

Table 10   Graph 6. Religions of the World

Religions

This Web site source claims that there are 4,300 different "faith groups" covering all the countries of the world. However, 98% of the world's population can be accounted for with a list of twenty-two religions. Included in these numbers are 16% who are "nonreligious" (atheist, agnostic, secular, humanistic, etc). All right, the twenty-two religions are shown on table 10 and graph 6. Actually, the top ten covers all but 2% of all religious groups. Note that Christianity is a catchall for the following: Catholic, Protestant, Eastern Orthodox, Pentecostal, Anglican, Monophysite, AICs, Latter-day Saints, Evangelical, Seventh Day Advocates, Jehovah's Witnesses, Quakers, Assemblies of God, etc.

Climatic Influences

We can go on forever with earth, but we must come to an end. However, one last subject bears mention—the ice ages. Back in the early 1960s, I studied basic and advanced geology at the University of Wisconsin. I distinctly remember the subject of ice ages and vaguely remember three or so ice ages in recent history. From recent research on the subject, it is now believed that in the last billion years the earth's climate has fluctuated between warm periods and cold periods on a somewhat periodic basis. On the extremes, a warm period might mean no ice at all, and an ice age might mean that glaciers covered most of the continental globe. There appears to be all kinds of in-betweens. Let's get the facts.

It starts with a dozen or so major glaciations in the Northern Hemisphere in the past one million years. The greatest ice event ever was about 650,000 years ago. Ice buildup was massive and extended deep into North America and Europe. The glaciers were so massive that the sea level dropped 400 ft, and the overall temperature across the globe dropped by 9°F. Recall that our ancestors, of species *Homo*, had arrived about 2 million years ago and had discovered fire from 1.5 million to 800,000 years ago. They most likely made their way to the Southern Hemisphere where hunting would have been good. This ice age lasted 50,000 years.

After this ice event, there was a succession of lesser ice ages roughly every 100,000 years since. The last ice age peaked roughly 20,000 years ago, perhaps in the later Cro-Magnon cave-dweller era. Sea levels rose in two major steps, one about 14,000 years ago, and the other about 11,500 years ago. But in between these, two meltdowns were another smaller ice age. The causes of these rather short-term swings in climate are not known!

But there are some theories. It is professed that small changes in the earth's orbit lead to climatically important change in the strength of seasons. This happens because the wobbles vary between a circular and an elongated orbit on a 100,000-year cycle, a pattern that coincides with ice ages over the past million years. To conceptualize this, understand that the earth in its orbit around the sun is tilted on its axis by 23.5°. If earth was not tilted, the sun would be directly over the equator all year long—one rotation around the sun. In the summer, earth's Northern Hemisphere is tilted toward the sun, and in the winter, it is tilted away from the sun. So in the height of summer (June 21), the sun appears directly overhead at northern latitude 23.5°, called the Tropic of Cancer. My globe shows this latitude cutting below all of the United States and all of Europe. It would cut just above Mexico City, slightly north of Cuba, through southern Egypt, through Saudi Arabia, India, and southern China. In the height of winter, the overhead sun would be 23.5° Southern Hemisphere. It would cut a line through Botswana, Africa, the island of Madagascar, Australia, Chile, Brazil, and others. Where the orbital "wobble" comes in, earth's tilt is not a steady 23.5°. Rather it wobbles between 22 to 24.5° over a cycle of 41,000 years.

Here is an interesting point of view. It argues that interglacial warming is most likely from heat from the earth's core because the oceans are heating up much more than the air, and that air temperature is overplayed in the hype of global warming. It claims that below-ground temperatures have been increasing. A lot of heat from the earth's core gets to the surface as evidenced by deep wells producing warm water. And with a rather controversial statement, "Increases in CO_2 in the atmosphere have been occurring for about a century. But only about 3% of the CO_2 has human origins. A likely source of the increase is heating of oceans, which causes CO_2 to be released." (*Science is Broken—Climate and Ice Ages, http://nov55.com/cli.html*).

I stand neither for nor against this argument. I bring it to your attention as a matter of interest.

Earth Summary

Earth was born as a massive fireball 4.5 billion years ago and has been an ever-changing entity ever since. It gradually cooled, at least to the point of sustaining water. Over a long time, water filled a giant reservoir, which was later called an ocean, and later yet oceans. A single massive piece of land was born and was called Rodinia. But later, Rodinia gave way through plate tectonics to the current landmasses that we see today as continents. The original atmosphere consisted of hydrogen, methane, carbon dioxide, ammonia, water vapor, and probably others but was devoid of oxygen. Life-forms began with this atmosphere (lacking oxygen), as simple one-celled creatures and gradually more complex life-forms, and then plants and sea animals. At some point, photosynthesis emerged, transforming the evolutionary landscape. This process developed the ability of plants to take energy from the sun to transform carbon dioxide into carbohydrates (cellulose) while emitting oxygen as a by-product. As oxygen made its presence in the atmosphere, the table was set for the emergence of land animals and human life-forms. And when life came, it flourished. But earth is not only ever-changing, it is a dangerous place. We learned about extinction events from volcanoes, ice ages, and collisions. But life was always able to renew its evolutionary process and come back with many other new species. And ultimately, human life with intelligence emerged. The dangers are still with us today. Volcanoes, ice ages, and collisions are still considered possible extinction events; add to this are infectious diseases, weapons of mass destruction, global warming, famine and population explosion, which are ever lurking and creating instability.

On the positive side, intelligence gives humankind a chance—an evolutionary advantage. We can develop medical cures, extend life expectancy, and develop agricultural techniques for proper distribution of the food chain. We can also, to a certain degree, anticipate disaster and fend it off or at least minimize the damages. We can study the universe under an umbrella called science, whose discipline calls for asking questions and strives to answer all of them. And perhaps one critical question is, how do we preserve the human species in the fate of certain death?

Literature indicates that the sun will deplete its fuel within 5 billion years at which time, of course, earth will cease to exist. Yet Charles Seife claims that our time on earth is a mere 1 billion years, probably indicating that earth will heat up to an unbearable cauldron long before the sun turns into a red star. So, folks, move your end of earth calendars up by 4 billion years and adjust your thinking. (*Alpha and Omega,* Charles Seife, p. 218).

Well, one billion years still gives science a lot of time to answer a whole lot of questions. By this time, we would have identified other habitable planets and conquered the intergalactic problems of space travel. If other intelligent life-forms are identified, we can only hope that they are friendly or weaker. That, of course, would preclude any earthly, cataclysmic, life-ending event. And lastly, assuming no other options, the human species will die out. And the universe will end anyway. Fatalistic? Perhaps! Reality? As of today, yes!

A press release on May 30, 2008, announced that astronomers had discovered the "smallest alien planet yet—"a super-Earth", only four times heavier than our home planet." Its distance to the nearest star puts it in a "habitable zone." If confirmed, it would be one of forty-five similar such planets recently found in the Milky Way by a European telescope located in Chile. It is now thought that these planets may be *very, very common.*" (Refer to Notes page for website.)

So maybe there is at least some hope. Listen to this story posted in *Astrobiology* magazine on April 10, 2008, and placed on a Web site (refer to Notes page). Its title is "Intelligence: A Rare Cosmic Commodity" by John D. Ruley. It deals with the improbability of finding intelligent life on other planets even though those planets would be "earth-like." Mr. Ruley writes that a Professor Andrew Watson developed an improved mathematical model for the evolution of intelligent life based on our experiences here on earth. Time wise, this would mean that 4.5 billion years would be required plus another 1 billion years more until the heat of the sun gets too hot for habitability. Professor Watson's premise is that approximately four major evolutionary steps would be required for this to happen: (1) single celled life-forms must evolve within a half-billion years after the birth of a planet, (2) multicellular life-forms must evolve about a billion and a half years after that, (3) specialized cells allow complex life-forms with functional organs about a billion years after that, and (4) and humans and human language must evolve about a billion years later. After certain assumptions, Watson concludes that the total probability that intelligent life would emerge is less than 0.01% over 4 billion years. But it must be said that the highly improbable does not mean the impossible.

The End of the Universe
The measuring tools for the age of the universe are in a constant state of improvement. The latest estimate is 13.73 billion +/- 120 million years, as per a press release on March 9, 2008. But it is time now to talk about the demise of the universe.

Let's go back to chapter 4, where I had promised you a surprise. And so it is. First, let's go way back to Einstein's belief in a static universe. Even when Hubble proved that the universe was expanding, Einstein stayed in denial. He even created the cosmological constant to try to disprove Hubble. But Hubble was right, and science began to study a beginning and an end. We saw that the big bang theory seems to satisfy many of the issues that other theories do not. In the 1980s, Guth resurrected Einstein's cosmological constant to help his understanding of his big bang numbers. In doing so, he recognized that the antigravitational forces far exceeded what Einstein ever envisioned. In fact, it accounted for the inflationary universe concept, which in turn explained the horizontal problem (review in chapter 4). Up to this point, scientists thought that the universe was expanding but was in deceleration. They discovered, totally unexpectedly, that the universe was not decelerating at all—in fact, it was accelerating, ever since 7 billion years after the big bang.

With this discovery, they turned the scientific community upside down. It would have been a bit more glamorous, perhaps, to know that the universe is slowing down; that it will stop some day. Then it will recede on itself until it became a point singularity. Then it would inflate again into a big bang event. Scientists already had a name for this scenario—a pulsating universe. But to accelerate into an ever-expanding, never-slowing universe until it cools and dies is hardly glamorous. It is devoid of hope and stunning to the faithful. It's like Clint Eastwood cleaning up a bad town and then forgetting to ride off into the sunset. It's like any movie that doesn't end right.

We can expect the fabric of the universe to continue expanding forever. Galaxies will recede from each another. Lost energy would result in a slower and colder universe. All the stars would burn out and die as the universe grows darker and darker. It will become an infinitely large, lonely void. It is referred to among the cosmologists as "The Big Chill."

Chapter Six

Strings—and Things

What is Superstring Theory

NOW THAT EARTH has died, and the universe is in purgatory for trillions and trillions of years before dying, we are going to go back to today, the year 2008. We had talked in previous chapters about scientists formulating theories and then challenging the same theories to make sure that they are right—or not right. When proven right, they become laws. And even laws are not sacred. They are laws until proven otherwise. Some theories cannot be proven either way, at least in the foreseeable future. Such is the case with string theory.

This subject, superstring theory or string theory for short, is an absolute eye-popper. For this subject, I will refer to Brian Greene's *The Elegant Universe*, as well as several Web sites.

String theory has found popularity over the last thirty years or so as a means to address mathematical quandaries such as the unification of gravity with the other three forces and dilemmas such as point singularities where density, temperature, and space-time curvature go out to mathematical infinity. Other dilemmas involve the two pillars of science, relativity and quantum mechanics, the mathematics of which rule, respectively, the macro (stars, galaxies and the universe itself) and the micro (atoms, atomic particles, and subatomic particles). Yet they are mutually incompatible! And lastly, the standard model gives no answers as to the details as to why the particles and the forces have the characteristics that they have been assigned to. Such dilemmas are extremely bothersome to mathematicians because they imply that something just ain't right. So they have continued at their "chalkboards" looking for new equations to free their minds of this mess.

In the 1960s and 1970s, particle accelerators were used widely to study the atomic strong force. Theories were being formulated to fit the information gathered. The big bang singularity was still a nightmare for mathematicians. One of two theories at the time was string theory. In 1968, a theoretical physicist by the name of Gabriele Veneziano, who was conducting studies at CERN in Switzerland, stumbled across some old mathematical formulas describing properties of strongly interacting particles. The formulas dated back about 200 years and were known as Euler's beta function. The formulas seemed to work, but no one knew why. Two years later, three scientists, Yoichiro Nambu, Holger Nielsen, and Leonard Susskind revealed the secret behind Euler's formula. They showed that if one modeled elementary particles as little, vibrating, one-dimensional strings, their nuclear interactions could be described exactly by Euler's equation. And the word *strings* was born. However, string theory had some bumpy roads ahead as some of the information from particle accelerators was very conflicting.

Some very positive news came in 1984. "Michael Green and John Schwarz established that the subtle quantum conflict afflicting string theory could be resolved and could take the string promise even further to include all four forces, and all of matter as well. As word of this result spread throughout the worldwide physics community, particle physicists by the hundreds dropped their research projects to launch a full-scale assault on what appeared to be the last theoretical battleground of the ancient quest to understand the deepest workings of the universe." Greene actually began his graduate work in 1983 at Oxford. He proudly continues that it was "an electrifying sense of being on the inside of a profound moment in the history of physics." (*The Elegant Universe*, Brian Greene, p.138)

This three-year period from 1984 to 1986 has now been called "the first superstring revolution." In this three-year period, there were more than one thousand research papers on the subject of superstrings written by physicists from across the world. Brian Greene quotes Michael Green as follows: "The moment you encounter string theory and realize that almost all of the major developments in physics over the last hundred years emerge—and emerge in such elegance—from such a simple starting point, you realize that this incredibly compelling theory is in a class of its own."

Nevertheless, the road to progress was filled with obstacles, detours, and hazards. In 1995, Edward Witten, in a presentation to a convention at the University of Southern California announced a plan for the next-forward step in string theory, thereby issuing in the "second superstring revolution." Witten proclaimed that "string theory is a piece of twenty-first century physics that happened to fall in the twentieth century." The problem is that the strings are so infinitesimally small that they cannot be "seen" and therefore the theory cannot be verified. Again, scientists are placing their faith in the purity of mathematics.

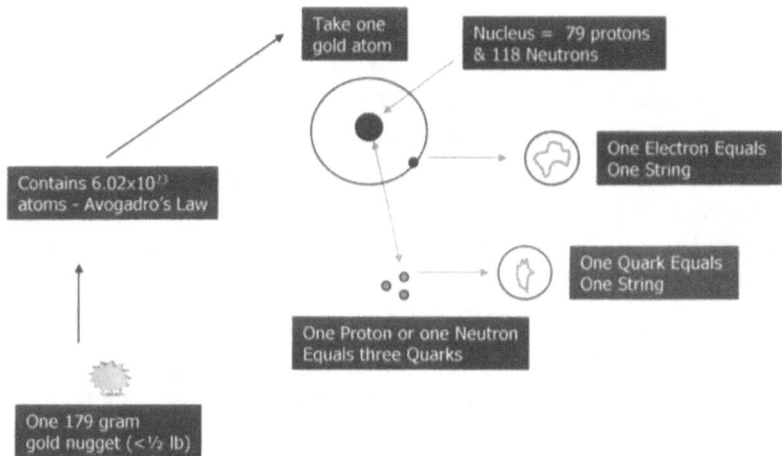

Illustration 13. Elemental Nature of Strings

So what is a string? It is two-dimensional, and it is elemental. Since an electron is a string, we can no longer think of it as spherical, ditto with quarks. Strings vibrate according to energy levels. Refer to illustration 13 for a visual. And hang on, strings are so small as to be infinitesimally incomprehensible.

First, I hope you are saying "two-dimensional—say what?" to the above comment. Not that I have an answer—I don't. I can only presume that the third dimension, depth, is so tiny as to be mathematically negligible.

Next, the string "size" issue has suddenly taken a leap into the twilight zone. Let's think this through. I am going back to my wall chart of fundamental particles and interactions, which shows a drawing of the structure within the atom. The "size" of the atom in this illustration is 10^{-10} meters. The size of the nucleus is 10^{-14} meters—10,000 times smaller than the atom. There are two protons and two neutrons shown (so it is a helium atom, right?) that occupy a space of 10^{-15} meters, 10 times smaller than the nucleus itself. There are twelve quarks shown, three per particle. Each quark is 10^{-19} meters. Each quark is 10,000 times smaller than the proton or neutron. So relative to the nucleus, one quark is 100,000 times smaller! OK, I can handle this so far.

Let's stay with the helium nucleus. There are two protons and two neutrons. Each has three quarks, for a total of twelve. Each is now a string, for a total of twelve strings. Now get this, a single string is one hundred billion billion (10^{-20}) times smaller than an atomic nucleus. In this example, that would mean 12×10^{-20} smaller than the nucleus. Now all of a sudden the string nucleus is infinitesimally smaller than the proton/neutron nucleus.

Could such small strings, three to each particle, generate the kind of mass to equal the mass of what we currently call our proton or neutron? If so, then the density of each string must be humongous. But if strings are two-dimensional, then they can't have a density for density is a measurement of volume. Is there that much energy implicit in three strings to achieve that kind of mass? There is nothing that I can draw that would demonstrate this magnitude of size difference, but I am going to try anyway. On illustration 14, I have drawn a string with a size of about 1/8 in. Next to it, I have drawn a sphere that is too big for the page. A quick calculation indicates that the bigger sphere is 11,000 times larger than the tiny string. Carrying this out further, another calculation indicates that the diameter of the larger sphere would have to be three hundred billion miles to simulate the difference between that 1/8 in single string in a nucleus. I have no recourse but to cast my faith in the hands of our very capable scientists, at least to the extent that it is backed up by the purity of mathematics.

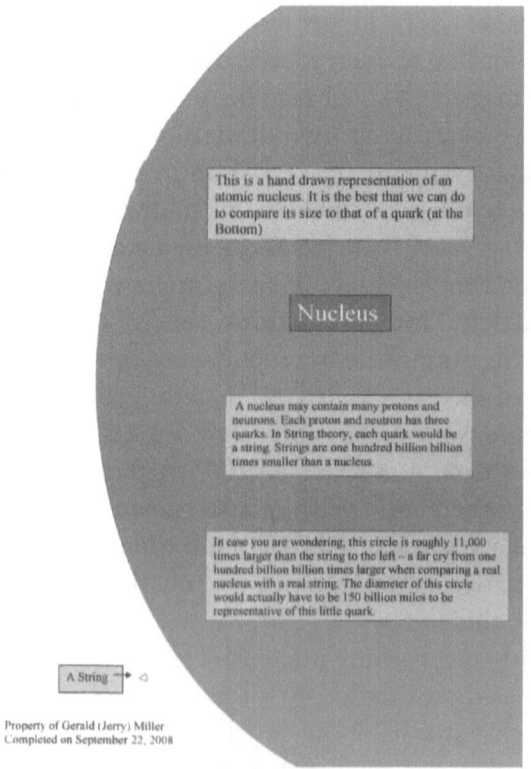

This is a hand drawn representation of an atomic nucleus. It is the best that we can do to compare its size to that of a quark (at the Bottom)

Nucleus

A nucleus may contain many protons and neutrons. Each proton and neutron has three quarks. In String theory, each quark would be a string. Strings are one hundred billion billion times smaller than a nucleus.

In case you are wondering, this circle is roughly 11,000 times larger than the string to the left – a far cry from one hundred billion billion times larger when comparing a real nucleus with a real string. The diameter of this circle would actually have to be 150 billion miles to be representative of this little quark.

A String

Property of Gerald (Jerry) Miller
Completed on September 22, 2008

**Illustration 14. Size Comparison
String vs Nucleus**

Now let's talk about string vibrations and its relationship to energy and mass. Greene reverts to real-life comparisons using actual string vibrations like those on a violin, piano, etc. Each string can resonate between point A and point B in an infinite amount of ways, resulting in a pattern of peaks and valleys. Our ears interpret these evenly spaced waves and valleys from a violin as music. Refer to wave illustrations on slide illustration 15. Each string resonates its own amplitude and trough in the form of energy. By Einstein's equation $E=mc^2$, energy increases as mass increases or vice-versa; the greater the amplitude and frequency, and the shorter the wavelength, the greater the energy. Now here is a crucial point—the energy of the string determines the mass of the particle, and the mass of the particle ties in directly to the gravitational force, as per $F = Gmm'/d^2$. This suggests a direct tie-in for the unification of the rogue gravity force with the other three pre-unified forces. It appears then that, similarly, the vibration of the string determines the particle charge, the weak charge and the strong charge of the particle. And the same holds true for the messenger particles—the photons, weak gauge bosons, and the gluons have their own vibrational pattern.

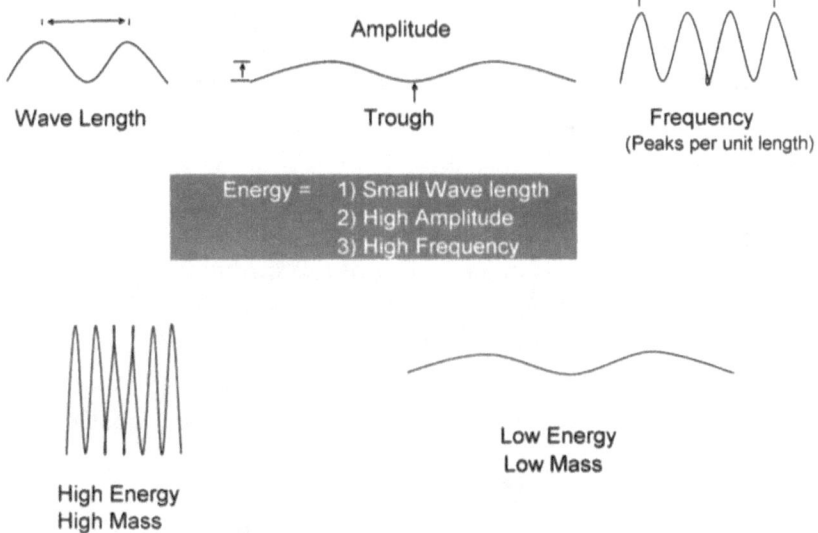

Illustration 15. Waves, Mass and Energy

This takes us back to the basic parameters of particles and forces discussed in chapter 2. What is so elemental about string theory is that all forms of matter and all forms of forces are all forms of strings. They are the same—they just vibrate *differently!* Energy changes with wavelength, frequency, and amplitude. As such their masses, charges, spins, and their flavors differ, but the strings are absolutely the same!

String theory today seems to have many advocates and many detractors. Feelings are strong on both sides as egos and careers are on the line. You have already heard the advocate's side.

Now from the detractors; an article was published on March 14, 2005 by Keay Davidson, Chronicle Science Writer which was subsequently posted on the internet. "Skeptics have long mocked string theory as untestable, because experimental studies of it would require machines of huge scale, perhaps even as big as the solar system." In another criticism "To critics . . . it is a disaster for string theory because the sheer number of estimated universes—equal to the number one followed by 500 zeros—is unimaginably large." *(Theory of Everything tying researchers up in knots, http://www.sfgate.com/cgi-bin/article.cgi?f=/c/a/2005/03/14/ MNGRMBOUREI.DTL&typ).*

And lastly—". . . recently theorists have estimated that there could be at least 10^{100} different solutions to the string equations, $10^{100}!$, corresponding to different ways of folding up the extra dimensions and filling them with fields—gazillions of different possible universes." (New York Times, December 7, 2004)

Despite the concerns about string theory it is not obviously going to disappear soon! After all, it took about 2,400 years to prove the existence of an atom.

M-Theory

The M-Theory is an advanced form of the simple string theory. String theory ran into a problem. Another version of the equations was discovered, then another, then another—well, five equations in all. They all appeared correct and all required ten dimensions. In the mid-1990s, not that long ago, Edward Witten and others suggested that all five equations might be saying the same thing with a different perspective. They proposed a unifying theory called the M-Theory, with M meaning membrane. It is often pictured as a five-point starfish. M-Theory brought all of the string theories together by theorizing that strings are really one-dimensional slices of a two-dimensional membrane vibrating in eleven-dimensional space. And more recently, another equation based on supergravity has been included to make a six-fingered starfish. See illustration 16.

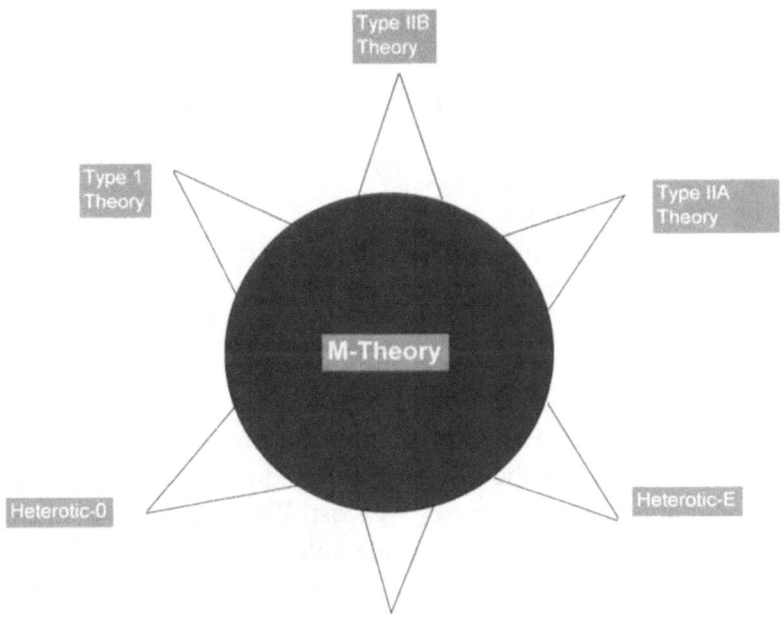

Illustration 16. M-Theory

So far, M-Theory is in agreement with all string theories as well as all scientific observations of the universe. Stephen Hawking and many other cosmologists have been attracted to the M-Theory for its mathematical elegance and relative simplicity. It is the most promising theory for the unification of the four forces.

... and Things

Antimatter

Antimatter particles are real, even though they sound like fiction. Shortly after the big bang, there should have been an equal number of matter and antimatter particles, as required by law—every particle has an antiparticle. Since a collision by matter and its antimatter results in total annihilation, why did matter win out on this exchange? Obviously, there is plenty of matter left over, and as far as scientists can tell not a whole lot of antimatter. Scientists do not know the answers, but one speculation is that more of the normal matter existed in the beginning of the universe. What was left over after annihilation is what makes up the visible universe. Speculation two is that normal matter was for some reason favored. In fact, matter and antimatter are not truly equal and opposite. The subtle differences must have favored normal matter. Whichever reason, we can be thankful that matter won.

Big Bang (refer to notes)

We know nothing about what was before zero time or what happened at zero time. Quantum mechanics has worked well in this arena, but only down to Planck time, which is 10^{-43} seconds. Physicists have already calculated that the weak, the strong, and the electromagnetic forces unified at somewhere prior to 10^{-35} seconds. In this state, the universe was far more symmetric than it is today. Under immense pressure and temperature, the forces are unified. As the universe cooled during expansion, the three forces disassembled, and symmetry was lost. We expect that in the future, mathematicians will be spending considerable time explaining the mysteries between zero big bang time and Planck time, 10^{-43} seconds. And indeed this is in the works.

Dark Energy

Speculations point to dark energy as an unknown force, perhaps Einstein's cosmological constant. In 1998, it was discovered that the universe's expansion was accelerating when it was thought to be decelerating. This dark energy is thought to contribute 65% to all the "stuff" that is out there. This form of antigravity seems to push galaxies apart thus neutralizing or overwhelming the force of gravity. This is very baffling and bewildering to cosmologists.

Dark Matter

Dark matter is sometimes called exotic dark matter that nobody has yet seen. Nobody knows in detail what the properties of dark matter are. Right now, what is known is that 30% of matter in the universe is dark. There is another 5% of ordinary matter, also called baryonic matter (protons and neutrons) that is observable.

Neutrinos

The search right now is for the missing 30% exotic matter. Quarks are not part of the search because they represent the 5% of visible matter, protons and neutrons. From what we know, now that leaves only the leptons for the rest of the mass. These include electrons, muons, and taus (refer to chapter

2). However, the muons and taus are heavy particles and are extremely unstable. They decay in millionths of a second and cannot be counted as part of the exotic matter. The electron is stable and commonly found in the universe but not as free electrons. The only particle left is the neutrino. Until recently, it has been very elusive, relative to mass and speed and of no charge. It is subject to intense scrutiny as this is published. You can expect to hear more on this subject in the coming years.

Symmetry and Supersymmetry

This is a subject that seems to intertwine the disciplines of philosophy, science, and mathematics with an abstract twist and symbolically glued together under the guise of aesthetics. Yet Brian Greene states that "symmetry in physics has a very concrete and precise meaning." He says that in developing theories, "use logical consistency,' 'avoid logical absurdities,' and 'have a sense of qualitative experimental implications of the theoretical construct relative to another.'" And he continues with "theoretical physicists are founded upon an aesthetic sense—a sense of which theories have elegance and a sense of beauty and structure on par with the world we experience." (*The Elegant Universe*, Brian Greene, p. 167-169)

OK! Let's try the concrete then. Physicists want the laws of nature to be solid, unchanging, forever, under all situations. Fortunately, physicists believe this to be so. Loosely defined, this is what physicists call *symmetry*.

By the mid-1970s, physicists realized that if the universe embraces supersymmetry, then every particle known must have a partner. One not as yet discovered. These partner particles would have a spin one-half lower than the spin of the original particle. Physicists actually have names for these particles such as sneutrinos, squarks, photinos, gluinos, winos, and zinos. However, physicists didn't welcome this turn of events since it added to the list of fundamental particles, muddying up the water. However, they still support symmetry for some reasons that are way beyond our scope. Look for progress in this area.

Multiple Universes

The thought of multiple universes would seem to me an uninvited distraction—just another in an eternity of surprising discoveries. Every time this story gets simple, it gets correspondingly complex. First, earth was once thought to be at the center of the world, without having any idea of what the world consisted of at the time. Which it wasn't, of course—it wasn't even the center of our solar system, which is only a tiny part of our galaxy, which is only one galaxy out of billions of galaxies. Many theories have been formulated. Some have graduated to the sanctity of constants and laws. Scientists have a good idea of how the universe began and how it will end. And now, we have to deal with multiple universes.

Vacuum

I read once that zero is the active presence of nothing, and nothing is the complete absence of anything. And indeed, a vacuum is the complete absence of anything—or is it? According to Charles Seife, physicists now believe that the vacuum holds the secret to the newest question in cosmology: what is this mysterious antigravity that pushes galaxies apart? But first let's focus on the vacuum. It was discovered as far back as the 1930s that a vacuum is not truly empty. It turns out that vacuums are filled with activity, with both particles and energy. OK! How so?

Much if not all of this is coming from quantum mechanics, which makes you want to run and hide to begin with. Well today, quantum physicists are forced to conclude that the vacuum is not truly empty. On larger scale, electrons and antielectrons are constantly popping in and out of existence. On smaller scale, larger particles like muon or tau particles become important. This is known as quantum vacuum. Several examples of an actual push of particles in a vacuum are reported. Metal plates have been moved! Hey! I'm convinced!

Get this! According to Charles Seife, "a vacuum is the leading contender for the source of dark energy, this elusive and so mysterious, pushing outward with such great intensity on all celestial bodies." *Alpha & Omega,* Charles Seife, p. 189)

This grouping of universal curiosities is by no means all inclusive. They were arbitrarily selected to let you know that scientists and mathematicians are continually creating and discovering and asking the big question *why?* They will not stop until all of the questions are answered, at which time we will know everything. After all, that is the purpose of science.

Chapter Seven

Game Changers of Science

Leucippus (490 BC-?): He was a philosopher from Greece credited with first considering the problem of splitting matter so small that it could no longer be split.

Democritus (460-370 B.): He was philosopher from Greece. Democritus is the first person to name this smallest piece of matter as "atomis" approximately 2,400 years before the discovery of the atom.

Ptolemy (70-174 AD): He invented a complex theory of planetary motion.

Nicolaus Copernicus (1473-1543): The first astronomer to propose the heliocentric arrangement of the planets, with the sun at the center and the other planets revolving around the sun—including Earth. Up until this point, earth was considered the center of the world. His book *De rovolutionibus orbium coelestium* was considered the beginning of modern astronomy and signaled the beginning of the Scientific Revolution. The book caused only mild religious controversy and provoked no fierce sermons for contradicting holy scripture. Three years later, however, a Dominican Catholic priest denounced his theory in the appendix of his book on scripture. Nicolaus was born in Poland and was an accomplished mathematician, astronomer, classical scholar, physician, translator, Catholic cleric, governor, military leader, diplomat, and economist. His discovery was nothing less than a scientific upheaval. The church didn't realize at the time that the first cosmological revolution had begun.

Galileo was born twenty-one years after the death of Copernicus. He was a physicist, astronomer, mathematician, and philosopher. He was called the "father of modern observational astronomy", "the father of modern physics" and "the father of science." His achievements include the improvements of the telescope and his resulting observations. With his telescope he was able to see things that no one had seen before and confirmed undeniably Copernicus' theory that earth was not the center of the world, but rather earth circled around the sun. Stephen Hawking "suggested that Galileo may have been the best scientist of the twentieth century." (*Stephen Hawking's Universe,* John Boslough), *and* (*http://en.Wikipedia.org/wiki/Galileo*).

The following is from Charles Seife, *Alpha & Omega*. Galileo was born into the midst of what is considered the first revolution of science. It was the first clash between science and religion. Before this revolution the cosmos belonged to theologians and philosophers. His telescope turned out to be the big weapon of the cosmological revolution. He became outspoken of his views, questioning the bible, bordering heresy. Galileo got called in by the head of the Roman Inquisition and was told not *to defend or hold the Copernican theory*. But he kept up with his ways. "In 1633, the Inquisition condemned Galileo as a heretic." His sentence was to be burned at the stake. He prudently recanted and was condemned to prison. As written by Seife, "as a favor to his old friend, Pope Urban VIII, allowed Galileo to spend his perpetual imprisonment at his home, rather than a dank cell in the Vatican." In 1822, the Catholic Church finally removed Copernicus's *On the Revolutions of the Heavenly Spheres,* Kepler's *New Astronomy,* and Galileo's *Dialogue Concerning the Two Chief Systems of the World* from the index of Forbidden Books. The church had accepted the new cosmology. (*Alpha & Omega,* Charles Seife, p. 22-23)

Sir Isaac Newton (1643-1727): According to Wikipedia he was an English mathematician, astronomer, physicist, philosopher and theologian. His publication in 1687, called the *Philosophiae Naturalis Principia Mathematica* is considered to be the most influential book in the history of science. In this book he described the laws of classical gravity and the three laws of motion. This laid the groundwork for classical mechanics and served as the basic view of the universe for the next three hundred years. The book is also the basis for modern engineering. His work on objects in motion helped remove any last doubts about earth and the planets rotating around the sun. In this regard, he helped advance the scientific revolution.

In mathematics, Newton is credited along with Gottfried Leibniz in developing calculus. (*http://en.wikipedia.org/wiki/Sir* Isaac Newton)

Thomas Young (1773-1829): He developed the wave theory of light.

Robert Boyle (1627-1691): With his J-tube mercury experiment, he gave strong evidence for the existence of atoms. It was a turning point in favor of atoms.

Michael Faraday (1791-1867): He developed the electric motor.

Christian Oersted: In 1819, he was the first to demonstrate a relationship between electricity and magnetism.

Andre Marie Ampere: In 1820, he discovered that electrical current in itself creates magnetism.

Michael Faraday: In 1831, he turned mechanical energy into electrical energy. His electromagnetic rotary devices formed the formation of electric motor devices.

Joseph Henry: In 1831, this American invented the first electric motor.

James Clerk Maxwell (1831-1879): He is known for color vision, molecular theory, and electromagnetic theory.

Joseph Thomson (1856-1940): In 1869, he discovered the electron and noted its negative charge.

Max Planck (1858-1947): In 1900, he suggested that radiation is quantized.

Dmitri Mendeleyev (1834-1907): In 1897, he is given credit for initiating the organization of the periodic table of the elements.

Albert Einstein (1879-1955): He proposed a photon that behaves like a particle/the equivalence of mass and energy/particle-wave duality/special relativity. In 1912, he explained the curvature of space-time.

Edwin Hubble (1889-1953): He was an American astronomer, who with the use of a phenomenon called the red shift profoundly changed our perception of the universe by demonstrating that there are other galaxies besides the Milky Way. He discovered that the degree of red shift observed in light coming from another galaxy increased in proportion to the distance of that galaxy to the Milky Way. In 1929, he proved that the universe was expanding, thus contradicting Einstein's steady state theory.

Hans Geiger, Ernest Marsden, and Ernest Rutherford: In 1909, they demonstrated the atoms to have a small, dense, positively charged nucleus.

Niels Bohr: In 1913, he constructed a theory of atomic structure based on quantum ideas. He also discovered that electrons occupy specific shells.

Ernest Rutherford: In 1911, he discovered that electros orbit an atomic nucleus. In 1919, he discovered evidence for the proton.

Werner Heisenberg: In 1927, he formulated the uncertainty principle.

Paul Dirac: In 1928, he combined quantum mechanics and special relativity to describe the electron. In 1931, he provided the first example of an antiparticle—a positron.

James Chadwick: In 1931, he discovered the neutron.

Hideki Yukawa: In 1933-34, he constructed a theory for new particles called pions—beginning of meson theory.

In 1948, the Berkeley synchrocyclotron produced the first pions.

In 1949, the K+ particle was discovered.

Donald Glaser: In 1952, he invented the "bubble chamber." The Brookhaven Cosmotron accelerator starts operation. 1953 was termed the beginning of the particle explosion.

Murray Gell-Mann and George Zweig: 1n 1964, they put together the theory of idea of quarks.

James Bjorken and Richard Feynman: In 1968-1969, their experiments at the Stanford Linear Accelerator provided evidence for quarks.

From 1970 to the present is mostly about new particles discovered in high-speed accelerators.

String theory was introduced in the early 1970s by Yoichiro Nambu, Holger Nielsen, and Leonard Susskind as a new theory for elemental particles. But it wasn't until 1984 when Michael Green and John Schwarz announced the possibility of unifying the four forces by using string theory, thus beginning of the first

In the 1990s, two teams of astronomers from Lawrence Berkeley and Australia discovered unexpectedly that the universe is currently in an accelerating expansion mode and has been so since 7 billion years after the big bang. The consequences of this finding are brutal. It means that the universe will end unceremoniously trillions of years from now as a dark ice cube.

The End

References

✓ Paul Halpern & Paul Wesson, *BRAVE NEW UNIVERSE*— ILLUMINATING THE DARKEST SECRETS OF THE COSMOS (2006)

✓ Brian Greene, *THE FABRIC OF THE COSMOS*—SPACE, TIME AND THE TEXTURE OF REALITY (2004)

✓ Charles Seife, *ALPHA & OMEGA*—The Search for the Beginning and End of the Universe (2003)

✓ Charles Seife, *ZERO*—The Biography of s Dangerous Idea (2000)

✓ Brian Greene, *The Elegant Universe*—Superstrings, Hidden Dimensions, and the Quest for the Ultimate Theory (1999)

✓ Matt Ridley, *GENOME*—THE AUTOBIOGRAPHY OF A SPECIES IN 23 CHAPTERS (1999)

✓ Isaac Asimov, *ATOM*—JOURNEY ACROSS THE SUBATOMIC COSMOS (1991)

✓ Steven Weinberg, *The First Three Minutes*—A Modern View of the Origin of the Universe (1977, 1988)

✓ 9) John Boslough, *STEPHEN HAWKING'S UNIVERSE*—An introduction to the most remarkable scientist of our time (1985)

✓ 10) Lincoln Barnett, *THE UNIVERSE AND DR. EINSTEIN*—A clear explanation of Einstein's theories and their effect upon the modern world (1948).

Notes

Chapter 1—The Basic Atom

Page 7, "Of the many Grecian philosophers . . ." Isaac Asimov, *Atom*, p 2-3.

Page 7, "In 1662, Boyle made use of a glass tube . . ." Isaac Asimov, *Atom*, p.5.

Page 8, "Elements" Isaac Asimov, *Atom*, p 6-8.

Page 10, "There are many versions of Periodic Tables . . ." and *http://www.webelements.com/webelements/scholar/*.

Page 10, "I have attached a listing of all of the elements . . . ." and *http://www.lenntech.com/Periodic-chart-elements-/alphabet.htm*.

Page 11, "The following discoveries finally expose . . . ." *http://particleadventure.org/particleadventure/other/history/earlyt.html*.

Page 12, "One real life mass comparison has it that . . . ." and *http://www.nyu.edu./pages/mathmol/textbook/atoms.html*.

Page 12, "Relative masses" *http://physics.nist.gov/cgi-bin/cuu/Value/mn*

Page 13, "Size of a nucleus" Isaac Asimov, *Atom*, p 96

Page 13, "If gold were to be rolled . . . ." Isaac Asimov, *Atom*, p. 94

Page 13, "How about a particle so small . . . ." Brian Greene, *Fabric of Space*, p. 346

Page 15, ". . . . within each element box is the actual electron code . . ."
http://www.chemicalelements.com/show/electronconfig.html.

Page 17, "Approximately 270 stable isotopes . . . ."
http://www.1bl.gov/abc/Basic.html.

Page 18, "By definition, radiation is the spontaneous . . . ."
http://www2.slac.stanford.edu/vvc/theory/nuclearstability.html

Page 20, ". . . in addition, there are non-metals . . . ."
http://www.chemicool.com/

Page 23, "In 1665, Isaac Newton . . . ." Isaac Asimov, *Atom*, p 27-28.

Page 24, "In 1819, a Danish physicist . . . ." Isaac Asimov, *Atom*, p. 42-43.

Page 24, "Magnetism"
http://maxwell.byu.edu/~spencerr/Phys442/node4.html

Page 25, "There are two more subjects . . . ."
http://en.wikipedia.org/wiki/Nuclear_fission.

Page 26, "Fusion is different than fission . . . ."
http://en.wikipedia.org/wiki/Nuclear.fusion.

Chapter 2—Subatomic Particles

Page 29, "First, let's have an historical perspective . . . ."
http://particleadventure.org/particleadventure/other/history/earlyt.html

Page 34, "Over the years . . . ."
http://science,howstuffworks.com/atom_smasher2.htm.

Pages 34-35, "All particle accelerators . . . ." Wikipedia—atom smashers

Page 36, "The number of accelerators . . . ."
http://www-elsa.physik.uni-bonn.de/accelerator_list.html

Chapter 3—Important Fundamentals in Physics

Page 41, "This force is actually . . . ."
http://en.wikipedia.org/wiki/weak.interaction,
http://en.wikipedia.org/wiki/strong.interaction.

Page 42, "Einstein's mass/energy equation . . . ."
http://en.wikipedia.org/wiki/Mass-energy_equivalence

Page 42-43, "Max Planck put forth . . . ."
http://en.wikipedia.org/wiki/Quantum_mechanics

Page 44, "If you remember nothing else . . . ." Charles Seife, *Alpha and Omega, p 43.*

Chapter 4—The Big Bang

Page 56, "Big Bang versus" Paul Halpern and Paul Wesson, *Brave New Universe,* pp86-92

Page 57, "Nebulas are cosmic occurrences . . . ."
http://seasky.org/cosmic/sky7a05.html.

Page 57, "Formation of Stars . . . .",
http://www.cs.csubak.edu/Physics/Phys110/UniverseScale.html.

Page 60, "Galaxies"
http://seds.org/Messier/move/mw.html, and
http://www.eso.org/public/outreach/press-rel/pr2004/pr-20-04.html.

Page 61, "The size of Galaxies"
http://seasky.org/cosmic/sky7a07.html.

Page 61, "Our Solar System"
http://en.wikipedia.org/wiki/solar_system.

Page 62, "This Chapter has taken us . . . ." Paul Halpern and Paul Wesson, *Brave New Universe,* Chapter 4.

Page 63, "The Horizon Problem" and "The Flatness problem" Brian Greene, *The Fabric of the Universe*, pp 294-303

Chapter 5—Birth of Earth, Evolution and End of World

Page 66-67, "Earth is Born"
http://en.wikipedia.org/wiki/History_of_earth, and
http://www.ecology.com/origins-of-life/earths_beginnings/index.html.

Page 67-68, "The Holocene Period"
http://www.extremescience.com/earth.htm

Page 69-71, "The Evolution of Life"
http://en.wikipedia.org/wiki/History_of_Earth

Page 72, "Human Chemistry"
http://chemistry.about.com/cs/howthingswork/f/blbodyelements.html

Page 73, "So, what is DNA?" Monroe Strickberger, *Genetics*, p. 49

Page 74, "When either of these two sugars . . . ."
http:/en.wikipedia.org/wiki/DNA

Page 78, "Below are milestone events . . . ."
http://en.wikipedia.org/Wiki/Hominid_intelligence

Page 80, "Earth and Humans—General Interest"
http://www.planetpale.com/planet2.html

Page 80, "Population and Trends"
http://www.geohive.com/default1.aspx

Page 81, "Wealth"
http://en.wikipedia.org/wiki/oil_reserves

Page 82, "Religions"
http://www.adherents.com/Religions_by_Adherents.html.

Page 83, "It starts with a dozen or so . . . ." *Climate Change, The Past and Future* *http://earthguide.used.edu/virtualmuseum/climatechange*

Page 83, "But there are some theories . . ." *http://www.koshlandscience.org/exhibitgcc/causes08.jsp.*

Page 84, "Here, is an interesting point of view" *http://nov55.com/cli.html.*

Page 85, "Literature indicates that the sun . . ." Charles Seife, *Alpha and Omega,* p. 218

Page 85, "So, maybe there is at least some hope." *http://www.space.com/scienceastronomy/080410-am-intelligence-model.html.*

Chapter 6—Strings and Things

Page 86, "Seife concludes . . . ." Charles Seife *Alpha and Omega*, p.219. And *http://www.time.com/times/covers/1101010625/story.html*

Page 94, "Antimatter particles are real . . . ." Charles Seife, *Alpha and Omega,* p. 131.

Page 94, "Pre-Zero time in Big Bang." Brian Greene, *The Elegant Universe,* P. 362

Page 95, "Symmetry and Supersymmetry," Brian Greene, *The Elegant Universe,* Chapter 7.

Page 96, "Vacuum" Charles Seife, *Alpha and Omega,* Chapter 12.

Chapter 7—Game Changers of Science

Page 98-102, "Game Changers . . . ." *http://particleadventure.org/particleadventure/other/history/early.html.*

Index

A

acceleration 56, 60-4, 72, 77
acceleration disk 72
adenine 89-90
AIC (African Inititated Churches) 98
alpha decay 25
Ampère, André Marie 34, 115
Andromeda (galaxy) 74
Anglicans 98
antigravity 68, 75, 110, 112
antimatter, see positive electrons
antineutrino 26, 40, 52
antiparticle 38, 41, 109, 116
antiproton 46
antiquark 43, 53
AOG (Assemblies of God) 98
Archean 80
Aristotle 14
Atom (Asimov) 16
   bomb 14, 18, 35
   clock 71
   mass 15, 18, 20, 24-7, 35
      units 25
   number 15-19, 25-9
   weight 15, 37
Avogadro, Amedeo 20
axions 75

B

base pair 87-8
Berkeley synchrocyclotron 117
beta decay 26-7, 40, 42, 52
big bang 29, 37, 46, 48, 65-9, 72-7, 79, 101-3, 109-10, 117, 123
binary star 72
Bjorken, James 117
black dwarf 70
Bohr, Niels 30, 54, 116
bosons 39, 42, 52, 107
Boyle, Robert 14, 115, 121
Brookhaven Cosmotron accelerator 117
bubble chamber 39, 45-6, 117

C

Cal Berkely, *See* University of California, Berkeley
Catholics 98, 113
CERN (Cernier Company) 104
Chadwick, James 18, 117
chain reaction 35
Chamberlain, Owen 46
Chao Chung Ting, Samuel 46
chromosome 86, 92
Clan of the Cave Bear (Auel) 83

cloud chamber 45

CMB (cosmic microwave background) 68

Copernicus, Nicolaus 113-14

Crick, Francis 88

cytosine 88

D

dark
    energy 75, 77-8, 110, 112
    matter 74-6, 78, 110

de Coulomb, Charles Augustin 33

Democritus 14, 113

dilations 58, 62-3

Dirac, Paul 38, 116

DNA (deoxyribonucleic acid) 85-8, 90-1, 124

E

Eastern Orthodox 98

Einstein, Albert 35, 38, 51, 54-6, 58, 60-2, 64-5, 75-6, 101, 116

electromagnetic
    field 51
    force 42, 50, 52
    radiation 26, 29-31, 35, 38, 42, 51-2

electromagnetism 48-50, 52, 54

electron 16, 18-19, 21-7, 29-30, 32, 35-6, 38, 40-2, 45-6, 52, 105, 111, 116

electron-neutrino 41

electron repulsion cloud 29

Elegant Universe, The (Greene) 55, 62, 103, 111

equivalence principle 61

Euler's beta function 104

Evangelical (religion) 98

exclusion principle 54

expanding universe 66

F

Faraday, Michael 115

Fermi National Accelerator (Fermilab) 44

Feynman, Richard 117

field 33-4, 39, 46, 51, 55, 71, 80, 92, 108. *See also* electromagnetic field; magnetic field

fission 34-7, 48

flatness problem 76

Fowler, William 67

Friedman, Alexander 66

Friedman, Jerome 46

Fuller, Buckminster 9

fusion 34, 36-7, 69-70

G

galaxies 48-9, 58, 66, 70-5, 103, 110, 112, 116, 123

Galilei, Galileo 113-14

gamma decay 27

gamma ray 26, 31

Geiger, Hans 18

Gell-Mann, Murray 39, 117

Genome (Ridley) 91, 119

Gilbert, William 33

Glaser, Donald 117

gluinos 111

gluons 39, 53, 107

gravitational force 34, 42, 49-50, 60-2, 107

graviton   42, 49, 52
gravity   32, 42, 48-53, 56-7, 60-2,
     64, 68-70, 72, 75, 77, 79, 103,
     107, 110, 114
  law of   49
Green, Michael   104
guanine   89
Guth, Alan   75

H

Hadean   79
Hawking, Stephen   92, 109, 114, 119
Heisenberg, Werner   54, 116. *See also*
     uncertainty
Henry, Joseph   34, 115
Holocene   80, 124
Homo sapiens   83
horizontal problem   76, 101
Hubble, Edwin   9, 65, 101, 116
hydrogen bonding   90-1

I

Indus River Valley   83
inflationary universe   101
"Intelligence: A Rare Cosmic
     Commodity" (Ruley)   101

J

J/psi particle   39
Jehovah's Witnesses   98

K

Kendall, Henry   46

L

Latter-day Saints   98
law of gravity   49
Lawrence Berkeley National
     Laboratory   77
Leibniz, Gottfried   114
leptons   38-9, 41-3, 46, 110
Leucippus   13-14, 18, 113
Lynn, James   92

M

M-Theory   108-9
MACHO (massive compact halo
     objects)   75
magnetic fields   34, 71
magnetism   33-4, 115
Magnus (Greek shepherd)   33
Mammoth Hunters, The (Auel)   83
mass   5, 15, 18-21, 35-6, 38-42, 49,
     54, 56-8, 74-7, 107, 116
Maxwell, James Clerk   54, 56, 115,
     122
MBR (microwave background
     radiation)   56, 76
Mendeleyev, Dimitri   15, 116
meson theory of nuclear forces   38
mesons   38-9, 41
Millikan, Robert   18
molecular weight   20
Monophysite   98
Moseley, Henry   15
muon   41-2, 112
muon-neutrino   41

N

Neanderthals 83
nebulas 70, 123
neutrinos 111
neutron 19-21, 24, 26, 30, 35-6, 38-41, 44, 46, 52-3, 105-6, 117
    beta decay 42
    star 70-2
Newlands, John 15
Newton, Isaac 31, 49, 114, 122
noble gases 27
nuclear power 35
nucleic acids 81
nucleosynthesis 67

O

Oersted, Hans Christian 34, 115

P

particle accelerators 44-7, 103-4
particle-wave duality 38, 116
Pauli, Wolfgang 38
Pentecostal (religion) 98
Penzias, Arno 68
Perl, Martin 46
Perrin, Jean Baptiste 19
photinos 111
photons 30-1, 38, 42, 52, 57, 67-9, 73, 107, 116
photosynthesis 80-2, 99
pions 42, 117
Planck
    Max 54, 116
    time 66
planets 7, 48-51, 58-9, 72, 74-5, 79-80, 100-1, 113-14

Plato 14
Plomin, Robert 92
positive electrons 45, 109-10, 125
primordial soup 7, 67
prism 31
Proterozoic 80
Protestant 98
proton 18-19, 24-6, 29-30, 32, 36, 38-41, 44, 46, 52-3, 105-6, 116
Ptolemy 113
pulsars 71
purines 89-90
pyrimidines 89-90

Q

Quakers 98
quantum
    hypothesis 54
    vacuum 112
quark 39-41, 46, 52-3, 105
    down 39-40, 52
    up 39-40, 52
quasars 71

R

radio waves 29, 31, 71
radioactive decay 26-7, 31
radioactivity 25, 40
red shift 71, 116
Reines, Frederick 46
Richter, Burton 46
Ridley, Matt 81, 91-3
RNA (ribonucleic acid) 85-6, 90-1
Rodinia 99
Rubbia, Carlo 46
Ruley, John D. 101
Rutherford, Ernest 18, 116

S

Schrödinger, Erwin 54
SDA (Seventh-Day Adventist) 98
Segre, Emillio 46
SLAC (Stanford Linear Accelerator in California) 44
sneutrinos 111
solid state electronic detectors 45
special theory of relativity 51, 56, 58, 60-2, 116
squarks 111
static electricity 29, 33
steady state theory 67, 116
strong force 42, 44, 48-9, 53, 103, 110
supercollider 46
supergiant 70-1
supernova 67, 69-71, 77
supersymmetry 111
symmetry 90, 110-11

T

tau 39, 41-2, 112
tau-neutrino 41
Taylor, Richard 46
Thompson, Joseph 116
Thomson, J. J. 18, 116
thymine 89
Townsend, John 18
trace elements 85

U

uncertainty principle 54, 116

V

vacuum 45, 61, 112
van der Meer, Simon 46
van Helmont, Jan Baptista 85
Veneziano, Gabriele 104

W

Walker, John 50
Watson
    Andrew 101
    James 88
wave theory 31, 38, 115
weak force 42, 48-9, 52-3, 110
weight 20, 25, 57, 85, 91, 94
white dwarf 9, 70, 77
white hole 9
Wilson, Robert 68
WIMPS (weakly interacting massive particles) 75
winos 111
Witten, Edward 104

Y

Young, Thomas 32, 38, 115
Yukawa, Hideki 117

Z

zinos 111